Plants of Central Asia

Plant collections from China and Mongolia

Volume 3

Plants of Central Asia

Plant collections from China and Mongolia

(*Editor-in-Chief*: V.I. Grubov)

Volume 3
Sedges–Rushes

T.V. Egorova

CRC Press is an imprint of the
Taylor & Francis Group, an **informa** business
A SCIENCE PUBLISHERS BOOK

ACADEMIA SCIENTIARUM URSS
INSTITUTUM BOTANICUM nomine V.L. KOMAROVII
PLANTAE ASIAE CENTRALIS

(secus materies Instituti botanici nomine V.L. Komarovii)

Fasciculus 3
Cyperaceae—Juncaceae

T.V. Egorova Confecit

First published 2000 by Science Publishers Inc.

Published 2019 by CRC Press
Taylor & Francis Group
6000 Broken Sound Parkway NW, Suite 300
Boca Raton, FL 33487-2742

First issued in paperback 2019

ISBN 13: 978-0-367-44749-6 (pbk)
ISBN 13: 978-1-57808-114-1 (hbk)
ISBN 13: 978-1-57808-062-5 (Set)

Library of Congress Cataloging-in-Publication Data
Rasteniia TSentral 'noi Azii. English
Plants of Central Asia: plant collections from China and Mongolia
/ [editor-in-chief. V.I. Grubov].
p. cm.
Research based on the collections of the V.L. Komarov Botanical Institute.
Includes bibliographical references
Contents: V. 3. Cyperaceae-Juncaceae
ISBN 1-57808-114-9 (v. 3)
1. Botany-Asia, Central. I. Grubov, V.I. II. Botanicheskii institut 1m. V.L. Komarova. III. Title.
QK374, R23613 1999 99-36729
581. 958-dc21 CIP

Translation of : Rasteniya Central'noy Asii, vol. 3, 1967
Nauka Publishers, Leningrad.
Addenda prepared by the author for the English edition in 2000.

NOTE

This volume represents the third in the series of illustrated lists of the plants of Central Asia (within the People's Republic of China and Mongolian People's Republic) published by the Botanical Institute, Academy of Sciences, USSR, based on the Central Asian collections of leading Russian travellers and naturalists (N. M. Przewalsky, G. N. Potanin and others) as well as by Soviet expeditions, and preserved in the Herbarium of the Institute. This volume is devoted to a description of the families of sedges, Araceae, duckweed and rushes. The members of the family of sedges (150 species) play a prominent role in the vegetation of Central Asia, especially in the higher altitudes, steppe and semidesert regions.

PREFACE

This volume deals with four families: Cyperaceae, Lemnaceae, Araceae and Juncaceae.

Two of these families, viz., Lemnaceae and Araceae, are poorly represented. They comprise only four species of hydrophytes and helophytes that are extremely common in the Northern Hemisphere and are of no particular interest.

Fam. Cyperaceae is represented by 13 genera and 182 species in the flora of Central Asia. One hundred and fifty species grow in Central Asia outside the USSR.* Of the 32 species not reported within this territory, 2 are found in Eastern Pamir and 30 in Kazakhstan. Further discussion deals with only the species found in Central Asia outside the Soviet Union.

Genus *Carex* with 96 species and *Kobresia* with 17 species, are the largest genera and are quite typical of high-altitude Central Asian flora.

Central Asian species of *Carex* are distributed among three subgenera—Primocarex (4 species), *Vignea* (19 species) and *Carex* (73 species)—and 23 sections. Species of three sections of subgenus *Carex* are most plentiful. These are: Atratae (14 species), Frigidae (11 species) and Lamprochlaenae (7 species). These 32 species constitute 35% of all *Carex* species inhabiting Central Asia outside the USSR. These three sections are distributed in temperate Eurasia and partly in North America. Members of the first two sections, viz., Atratae and Frigidae, grow predominantly in alpine or actic-alpine regions while those of section Lamprochlaenae grow in steppe expanses.

Among the species of *Carex* of Central Asia, quite a number may be reckoned as typically Central Asian given the nature of their present distribution. Such species number 21 or 22% of the *Carex* species in this region. From subgenus *Vignea,* they include C. *pycnostachya* (section Holarrhenae), C. *stenophylloides* (section Boernera) and C. *roborowskii* and C. *pseudofoetida* (section Foetidae). All these sections are distributed in Eurasia and North America and bring together first of all the boreal species and species of predominantly lower and middle montane belts; second, the steppe and semidesert species in the plains and hills; and third, the predominantly high-altitude marsh-meadow species. The closest genetic affinities are seen between C. *pycnostachya* and southern Siberian hill species C. *curaica;*

***As this book was published in Russia in 1967, the erstwhile abbreviation 'USSR' is retained (rather than the current abbreviation 'CIS')—General Editor (of the English edition).**

C. stenophylloides and steppe species European *C. stenophylla* Wahl. and Mongolian-Siberian *C. duriuscula; C. roborowskii* and Siberian hill-meadow species *C. enervis;* and finally *C. pseudofoetida* and European alpine plant *C. foetida* All. The typical Central Asian species of subgenus *Carex* are *C. melanantha* and *C. moorcroftii* (section Atratae), *C. aridula, C. ivanoviae, C. minutiscabra, C. relaxa* and *C. turkestanica* (section Lamprochlaenae), *C. alajica, C. alexeenkoana, C. montis-everesti, C. przewalskii* and *C. stenocarpa* (section Frigidae), *C. orbicularis* (section Acutae), *C. titovii* (section Mitratae), *C. aneurocarpa* and *C. lanceolata* (section Digitatae) and *C. pamirensis* (section Physocarpae). Mention has already been made of the first three sections (Atratae, Lamprochlaenae and Frigidae) which account for the largest number of species. The rest of the foregoing sections are boreal and extensively distributed in Eurasia and North America.

The more interesting phytogeographic contacts among the above species are as follows. *C. moorcroftii* has close genetic affinity with the Mongolian-Southern Siberian *C. sabulosa* Turcz. ex Kunth. *C. alajica* is relatively close to the Caucasian species *C. pontica* Albov. *C. stenocarpa* is extremely close to the Caucasian *C. tristis* M.B. and Central European *C. sempervirens* Vill. *C. pamirensis* is rather poorly isolated from the Mongolian-Daurian species *C. dichroa. C. aneurocarpa* bears genetic affinity with the Euro-Siberian *C. pediformis. C. orbicularis,* belonging to the largest and most widespread section of subgenus *Carex,* viz., *Acutae,* has close genetic affinity with the Altay *C. altaica* Gorodk. and the Caucasian *C. kotschyana* Boiss. et Hohen. and less close association with artic and bald-peak species of the group Rigidae [*C. ensifolia* (Turcz. ex Gorodk.) V. Krecz. and *C. bigelowii* Torr. ex Schwein.]. It is interesting that section Acutae, with rich representation in the arctic zone (20 species) has only six species in the territory of our study. Further, except for *C. orbicularis,* none of these species extends beyond its boundaries.

Five species of *Carex* are endemic to Central Asia. High-altitude species *C. montis-everesti* (South. Tibet) and *C. przewalskii* (Nanshan) growing in the middle to alpine montane belt are endemic to the Tibetan province. Two species inhabit Junggar: hill-steppe species *C. titovii* (Junggar Alatau and Tien Shan) and forest species *C. minutiscabra* (Tien Shan). Finally, the fifth endemic species, *C. aridula,* is found in Mongolia (Alashan mountain range) and Tibet (Nanshan and Weitzan). All these species have an extremely narrow distributional range confined to just a few sites. *C. montis-everesti, C. przewalskii* and *C. minutiscabra* represent morphologically distinct species. The first of these three occupies a fairly isolated position in section Frigidae, the second has genetic affinity with the Himalayan species *C. haematostoma* and the Siberian *C. macrogyna,* while the third is close to the Central Asian *C. turkestanica*. Insofar as the other two endemic species are concerned, viz., *C. titovii* and *C. aridula,* they are very close to the Siberian-European *C. caryophyllea* and Siberian *C. korshinskyi* respectively.

Some typical Central Asian species are very widely distributed in the Central Asian territory and are found in all its provinces. These are the high-altitude (predominantly) species *C. pseudofoetida, C. orbicularis, C. melanantha, C. alexeenkoana* and *C. stenocarpa* and the desert-steppe species *C. stenophylloides* found from submontane planes to high-altitude regions. The rest of the species are typical of one or two Central Asian provinces. Thus, species of high-altitude and moderate level hill sands—*C. moorcroftii* and *C. ivanoviae*—are characteristic of the Tibetan province. The hill-steppe species *C. relaxa* is confined to the Mongolian subprovince.

Other species—*C. turkestanica, C. aneurocarpa* and *C. pycnostachya*—grow mainly in Junggar-Turan province. The first two are hill-steppe species and the third—*C. pycnostachya*—a marsh-meadow plant usually inhabiting foothills to mid-montane belt regions.

Most typical Central Asian species of *Carex* belong to sections more advanced in evolution. Only *C. pycnostachya* and *C. pamirensis* belong to relatively old groups.

The ecology of *Carex* species discussed so far is diverse. They are found at all high levels but predominantly in the high-altitude regions, and are represented in different types of vegetation, ranging from meadow and marsh to steppe and semidesert. Some species are found in large numbers, forming nearly pure associations. Characteristic species of high-altitude marshy sedge meadows are: *C. pseudofoetida, C. orbicularis, C. stenocarpa* and *C. melanantha.* In the sandy high-altitude regions of Tibet, *C. moorcroftii* is found in abundance and constitutes an excellent sand binder.

Boreal species represent a large group of Central Asian species of *Carex* and play a very significant role in the plant cover (*C. lithophila, C. canescens, C. vulpina, C. pallida, C. obtusata, C. appendiculata, C. cespitosa, C. fuscovaginata, C. schmidtii, C. panicea, C. sedakowii, C. rhynchophysa, C. vesicata, C. rostrata, C. pseudocyperus, C. drymophila, C. orthostachys* and others). Most of these species are found on the northern and eastern boundaries of Central Asia. Some other species, of relict nature, are also present (*C. lithophila*—Mongolian Altay, *C. vulpina*—Altay district of Junggar and northern Junggar Gobi, and *C. canescens*—Altay district of Junggar and Tien Shan).

The following species represent arctic-alpine elements in the flora of Central Asia: *C. microglochin* (incidental reports in Hobdo district, Tien Shan, Nanshan and Weitzan), *C. rupestris* (Hobdo district and Gobi Altay), *C. tripartita* (Altay district of Junggar), *C. norvegica* (Hobdo district), *C. ledebouriana* (Hobdo district and Mongolian Altay) and *C. atrofusca.* The last species is highly characteristic of the *Carex-Kobresia* high-altitude meadows of Tien Shan and the Tibetan province.

The following Siberian steppe species are extensively and massively distributed within the Mongolian subprovince: *C. duriuscula* (entire Mongolia except the southern districts), *C. korshinskyi* (northern and eastern Mongolia) and *C. pediformis* (predominantly Mongolian Altay and East. Mongolia).

Siberian bald-peak species forming a distributional range in Central Asia cut off from the main range are probably relics of the Glacial epoch. These species are: *C. curaica* (Mongolian Altay and Altay district of Junggar), *C. macrogyna* (Gobi Altay) and *C. perfusca* (Junggar Alatau and Tien Shan).

Alpine Himalayan species penetrating into Tibet or into Pamir form yet another group of Central Asian species of *Carex* (*C. infuscata, C. pseudobicolor, C. duthiei* and *C. nivalis*).

Finally, southern Chinese species comprise an extremely small group of *Carex* species: alpine species *C. kansuensis* and forest species *C. lehmannii*. These species are found in Qinghai.

As mentioned above, genus *Kobresia* is represented in Central Asia by 17 species, which constitutes half of all its known species. The rest of the species inhabit the Himalayas, South-western China and North America. Typical Central Asian *Kobresia* species number eight: *K. capilliformis, K. smirnovii, K. tibetica, K. ovczinnikovii, K. robusta, K. humilis, K. pusilla* and *K. stenocarpa*. Almost all these species, except for *K. pusilla* (a rare and dubious species), play a major role in the formation of the plant cover of Central Asian high altitudes. *K. capilliformis* and *K. ovczinnikovii*, while being mutually close, show genetic affinities with the Himalayan *K. capillifolia*. *K. capilliformis* is the most characteristic plant of alpine sedge-kobresia meadows and is occasionally found even in the forest belt. This species mainly inhabits Tien Shan and Pamir. *K. ovczinnikovii*, while evidently widely distributed in Eastern Pamir, is known in a few sites outside. *K. smirnovii*, inhabiting marshy meadows and rubble placers in the alpine belt of Mongolian Altay and Junggar (Altay district and Tarbagatai) has very close genetic affinity with the Caucasian-Asia Minor species *K. schoenoides* (C. A. Mey.) Steud. *K. smirnovii* is also close to Pamiroalay *K. pamiroalaica* Ivanova and to *K. tibetica*, the latter two species being mutually close. *K. pamiroalaica*, a very common plant of Pamiroalay, especially Pamir, is not found in Central Asia outside the Soviet Union. *K. tibetica* inhabits the high-altitude marshes and marshy meadows in the Tibetan province. *K. robusta*, which occupies an isolated place in the genus because of the presence of closed perigyniums, is a highly typical plant. Its ecology is also interersting. *K. robusta* grows on high-altitude sand in the Tibetan province (except Pamir) and Qaidam. *K. stenocarpa* growing on wet sedge-kobresia meadows, less often on steppe slopes, is quite widely distributed in the high altitudes of Junggar (Junggar Alatau and Tien Shan), Kashgar and the Tibetan province. This species is very close to the Himalayan *K. royleana*. *K. stenocarpa* shows less genetic affinity with *K. humilis*. The latter is common in Junggar Alatau, Tien Shan and partly Nor. Kashgar.

K. humilis is more characteristic of moderate- (wet meadows and steppe-covered slopes) than high-altitude belts.

Subcircumpolar arctic-alpine species *K. bellardii* is widely distributed in some regions of Central Asia. Its habitats—meadows, hill steppes, turf-covered placers—are confined to alpine and subalpine, and less often forest belts. This species is more characteristic of Mongolian and Gobi Altay and less so of high-altitude Junggar. Two reports of this species, set off from the main distribution range, are known from Qinghai (Nanshan).

Another subcircumpolar arctic-alpine inhabitant, *K. simpliciuscula,* is extremely scarce within this subregion. There are two reports of it in Mongolian Altay and one in Gobi Altay. The Siberian-Mongolian species *K. filifolia* is found more often in Mongolian Altay. It is confined to meadow-steppe groups of forest and subalpine belts. *K. filifolia,* like *K. bellardii,* was found, after a long break, in Qinghai and, additionally, in North-western China.

High-altitude Himalayan species *K. capillifolia* and *K. royleana* as well as *K. kansuensis* enter Central Asia (Tibet). The main distribution range of the last of the aforesaid species falls in North-western and South-western China.

Representatives of other genera of fam. Cyperaceae (*Cyperus, Juncellus, Pycreus, Scirpus, Bolboschoenus* etc.) are, by and large, rather scant in Central Asia. They grow in lowlands and foothills of a low montane belt, extending occasionally to higher levels. Only species of genus *Trichophorum* and partly species of genus *Eriophorum* inhabit high hills. It should be noted that cotton grasses (5 species) are known only in a few stray sites falling on the northern and eastern boundaries of the subregion. The following species are quite representative of Central Asia: *Trichophorum distigmaticum* (Nanshan and Weitzan), *Scirpus hippolyti* (Mongolian province, Junggar), *Bolboschoenus planiculmis* (Mongolia, Kashgar, Junggar and Qinghai) and *Blysmus sinocompressus* (Mongolian province, Junggar, Qinghai and Pamir). Among these species, only *Trichophorum distigmaticum* grows on marshy grass plots along river banks in the high-altitude belt. The rest of these species inhabit solonchak bogs, swampy meadows and grass plots, banks of water reservoirs—from lowlands to the lower, less often middle montane belts.

Fam. Juncaceae is represented in the Central Asian flora by 25 species of genus *Juncus* and three species of genus *Luzula.* Their role in the plant cover is comparatively less important. These genera comprise no species specific for, let alone endemic to, Central Asia.

The most numerous group of rushes (9 species), subgenus *Stygiopsis,* covers alpine and arctic-alpine, predominantly Asian species. These are rushes exhibiting an affinity with J. triglumis, namely *J. allioides, J. potaninii, J. przewalskii, J. tibeticus, J. thomsonii* and others. All these species inhabit high-altitude Tibet but reports of their occurrence are few. Other distributional ranges include the Himalayas (*J. membranaceus*) or the

Himalayas and hills of North-western, Central and South-western China (*J. thomsonii, J. leucomelas* and *J. sphacellatus*) or exclusively the aforementioned hilly regions of China (*J. allioides, J. potaninii, J. przewalskii* and *J. tibeticus*). *J. triglumis* is an arctic-alpine species found within Central Asia, besides Tibet, even in Mongolia (one record each in Hobdo district and Mongolian Altay) and Junggar (Junggar Alatau and Tien Shan). Its usual habitats are wet and marshy grass plots along rivers and brooks, swampy meadows predominantly in the alpine belt and less frequently in the upper part of the forest belt.

The rushes commonly found in some regions of Central Asia are *J. salsuginosus* (Mongolia), *J. gerardii* (Mongolian province and Junggar) and *J. articulatus* (Mongolian province, except Qaidam, and Junggar). The first of these species is Mongolian, inhabiting hilly solonchak swamp meadows. The other two species are distributed mainly in the forest zone of Eurasia. Their ecology is diverse—solonchak-like swamp meadows, solonchaks, grassy bogs, pebble beds and silty-sandy sites along banks of water reservoirs. They inhabit submontane flats, foothills and lower hill belts, extending occasionally up to 2800 m above sea level.

Of the three species of genus *Luzula* reported in Central Asia, two, i.e., *L. spicata* and *L. sibirica*, are found in the alpine and subalpine zones of Mongolia and Junggar and the third, *L. pallescens*, in the forest and subalpine levels of Junggar Alatau. The main distribution range of these species falls outside the boundaries of Central Asia. *L. spicata* is a boreal Eurasian species while *L. sibirica* is a hilly, predominantly Siberian-Mongolian one.

Artist T. N. Shishlova did the drawings in the Plates presented in this volume.

The author expresses his gratitude to O. I. Starikova who was very helpful in processing the herbarium material of Chinese collectors (translation of labels and location of several geographical sites on the map).

CONTENTS

TAXONOMY

SPECIAL ABBREVIATIONS

Abbreviations of Names of Collectors

Bar.	— V. I. Baranov
Chaff.	— J. Chaffanjon
Chaney	— R. W. Chaney
Czet.	— S. S. Czetyrkin
Divn.	— D. A. Divnogorskaya
Fet.	— A. M. Fetisov
Glag.	— S. A. Glagolev
Gr.-Grzh.	— G. E. Grum-Grzhimailo
Grub.	— V. I. Grubov
Gus.	— V. A. Gusev
Ik.-Gal.	— N. P. Ikonnikov-Galitzkij
Ivan.	— A. F. Ivanov
Kal.	— A. V. Kalinina
Kashk.	— V. A. Kashkarov
Klem.	— E. N. Klements
Kondr.	— S. A. Kondrat'ev
Krasch.	— H. Krascheninnikov
Kryl.	— P. N. Krylov
Lad.	— V. F. Ladygin
Lavr.	— E. M. Lavrenko
Lee, Lee and Chu, Lee et al.	— A. R. Lee (1959)
Li, Sh. S. et al.	— S. H. Li et al.
Lis.	— V. I. Lisovskii
Litv.	— D. Litvinov
Mois.	— V. S. Moiseenko
Pal.	— I. V. Palibin
Pavl.	— N. V. Pavlov
Petr.	— M. P. Petrov
Pob.	— E. G. Pobedimova
Pop.	— M. G. Popov
Pot.	— G. N. Potanin

Przew.	— N. M. Przewalsky
Reg.	— A. Regel
Rhins	— J. L. Dutreuil de Rhins
Rob.	— V. I. Roborowsky
Sap.	— V. V. Sapozhnikov
Serp.	— V. M. Serpukhov
Shishk.	— B. K. Shishkin
Sold.	— V. V. Soldatov
Tug.	— A. Ya. Tugarinov
Yun.	— A. A. Yunatov
Zam.	— B. M. Zamatkinov

Abbreviated Names of Herbaria

BM	— British Museum of Natural History: London, Great Britain
K	— The Herbarium, Royal Botanic Gardens: Kew, Surrey, Great Britain
Linn.	— The Linnean Society of London: London, Great Britain

Family **CYPERACEAE** Juss.

1. Plant dioecious ....12. **Kobresia** Willd. (*K. prattii* Clarke).
+ Plant monoecious ....2.
2. Flowers unisexual, without perianth; pistillate flower and later even fruit covered by modified bractlet, closed or not; bractlet sessile in axil of glume ....3.
+ Flowers bisexual, with perianth comprising bristles or hairs or without perianth; modified bract absent; flowers directly in axil of glumes ....4.
3. Bractlet surrounding pistillate flower and fruit closed, in form of utricle. Stem largely foliate..13. **Carex** L.
+ Bractlet surrounding pistillate flower and fruit not closed or closed only at base, less often up to half or more. Stem foliate only at base ....12. **Kobresia** Willd.
4. Flowers with glumes forming distichously arranged spikelets; latter rather flat; perianth absent..5.
+ Flowers with glumes forming spirally arranged spikelets in many rows; spikelets rounded in section; perianth generally present....... 8.
5. Stigmas three; nut trigonous..1. **Cyperus** L.
+ Stigmas two; nut flattened, biconvex...6.

6. Nut with rib turned towards axis of spikelet.....4. **Pycreus** Beauv.
+ Nut with face (planar side) turned towards axis of spikelet.......7.
7. Inflorescence capitate, up to 1 cm long, spikelets in inflorescence not more than 10. Annual plant ....3. **Acorellus** Palla.
+ Inflorescence umbellate-paniculate, up to 7 cm long, with many spikelets. Perennial plant...................2. **Juncellus** (Griseb.) Clarke.
8(4) Spikelets distichous, forming compound spike at apex of stem. Stigmas two ....9. **Blysmus** Panz.
+ Spikelets in compound umbellate, paniculate or capitate inflorescences, or spikelets single..9.
9. Spikelet single..10.
+ Spikelets few or many...11.
10. Leaf sheaths invariably without blade. Style usually with greatly thickened base, stylopodium, seen in fruits...11. **Eleocharis** R. Br.
+ Upper leaf sheaths with short blade. Style base not thickened...... ....6. **Trichophorum** Pers.
11. Perianth bristles numerous and, after flowering, greatly elongate into silky hairs, considerably longer than glumes, and forming downy heads above spikelets...............................5. **Eriophorum** L.
+ Perianth bristles 4 to 6, short, not longer than glumes; bristles sometimes absent...12.
12. Spikelets small, not longer than 2.5 mm, forming 1 to 3 dense spherical heads 3 to 6 mm in diameter. Perianth bristles absent.. ....10. **Dichostylis** Beauv.
+ Spikelets considerably large; if small, not forming heads. Perianth bristles generally developed...13.
13. Inflorescence apparently lateral since lower bract leaf appears to be continuation of stem. Leaves generally seen as scaly sheaths.. ....7. **Scirpus** L.
+ Inflorescence distinctly terminal. Leaves flat ....14.
14. Spikelets few, brownish or pale yellow, large, 1 to 2 cm long. Rhizome often with globose thickenings, 'tubers'.. ....8. **Bolboschoenus** (Aschers.) Palla.
+ Spikelets numerous, greenish, small, 3 to 8 mm long. Rhizome without globose thickenings ....7. **Scirpus** L.

1. **Cyperus** L.

Sp. pl. (1753) 44; id em, gen.pl., ed. 5 (1754) 26.

1. Spikelets forming very dense ovate, spherical or oblong heads. Nuts oblong, thrice longer than broad.............2. **C. glomeratus** L.
+ Spikelets in rather lax fascicles. Nuts elliptical or obovate, twice longer than broad ....1. **C. fuscus** L.

1. **C. fuscus** L. Sp. pl. (1753) 46; Forbes and Hemsley, Index Fl. Sin. 3 (1905) 212; Krylov, Fl. Zap. Sib. 3 (1929) 380; Shishkin in Fl. SSSR, 3 (1935)

15; Kük in Pflanzenreich, [IV, 20], 101 (1936) 235; Persson in Bot. Notis. (1938) 275; Kitag. Lin. Fl. Mansh. (1939) 114; Fl. Kirgiz. 2 (1950) 239; Grubov, Consp. fl. MNR (1955) 79; Fl. Kazakhst. 2 (1958) 7; L. K. Dai in Fl. R. P. Sin. 11 (1961) 150; Fl. Tadzh. 2 (1963) 20. —**Ic.**: Fl. SSSR, 3, Plate 1, fig. 1.

Described from Europe. Type in London (Linn.).

Along moist sandy and silty banks of rivers and lakes and among swamps; in the lower hill belt, up to 1350 m above sea level.

IA. Mongolia: ***Cent. Khalkha*** (Borokhchin river, silty bank, 3 IX 1926—Ik.-Gal.), ***East. Mong.*** (right bank of Huang He river below Hekou town, on sandy-silty moist soil, 4 VIII 1884—Pot.), ***Bas. Lakes*** (Shargin Gobi, Shargin Gol river bank, on moist sand, 8 IX 1930—Pob.), ***Ordos*** (Dzhasygen-Qaidam area, 30 VIII 1884—Pot.).

IB. Kashgar: ***West.*** (Yangishar, 25 VIII 1929—Pop.; 'Kashgar, in a ditch on saline soil, nearly 1330 m, 17 VIII 1933'—Persson, l.c.), ***East.*** (south. fringe of Khami oasis, Bugas region, 480 m, along springs and near water 18 and 20 VIII 1895—Rob.; Turfan, Lake Yaer, near water, 100 m, No. 5563, 9 VI 1958—Lee and Chu).

IIA. Junggar: ***Altay region*** (on Qinhe river, 500 m, 11 IX 1956—Ching), ***Tien Shan*** (Ili-Chapchal, on weakly brackish soil, No. 3069, 5 VIII 1957—Kuan), ***Dzhark.*** (Kul'dzha, 5 VII 1877—A. Reg.), ***Balkh.-Alak.*** (around Dachen, in swamp, No. 2843, 10 VIII 1957—Kuan).

General distribution: Aral-Casp., Balkh. region, Jung.-Tarb., Nor. and Cent. Tien Shan; Europe, Mediterr., Balk.-Asia Minor, Near East, Caucasus, Mid. Asia, West. Sib. (south, excluding Altay), East. Sib. (south., excluding Sayans, rarer), China (Dunbei, North., North.-west., Centr.), Himalayas (Kashmir), Indo-Malay., North. Amer. (east.).

Note: In "Flora of China" (L. K. Dai, l.c.), this species has been cited for Inner Mongolia with no precise location indicated.

2. C. glomeratus L. Cent. pl. 2 (1756) 5; Franch Pl. David. 1 (1884) 316; Forbes and Hemsley, Index Fl. Sin. 3 (1905) 213; Krylov, Fl. Zap. Sib. 3 (1929) 381; Schischkin in Fl. SSSR, 3 (1935) 18; Kük. in Pflanzenreich [IV, 20], 101 (1936) 147; Kitag. Lin. Fl. Mansh. (1939) 114; Fl. Kazakhst. 2 (1958) 8; L. K. Dai in Fl. R. P. Sin. 11 (1961) 140.

Described from Europe. Type in London (Linn.).

IA. Mongolia: ***East. Mong.*** (Khailar, in steppe and on lowlands, 1959—Ivan.).

IIA. Junggar: ***Jung. Gobi*** (south.: Takienzi, 300 m, 6 and 24 VIII 1878—A. Reg.).

General distribution: Balkh. region, Jung.-Tarb.; Europe (south), Caucasus, Mid. Asia (nor.), West. Sib. (south., excluding Altay), Far East (south.), China (Dunbei, North., North.-west., Centr.), Himalayas (Kashmir), Korean peninsula.

2. Juncellus (Griseb.) Clarke

in Hook. f. Fl. Brit. Ind. 6 (1893) 594. —*Cyperus* L. sect. *Juncellus* Griseb. Fl. Brit. West-Ind. Isl. (1864) 562.

1. J. serotinus (Rottb.) Clarke in Hook. f. Fl. Brit. Ind. 6 (1893) 594; Forbes and Hemsley, Index Fl. Sin. 3 (1905) 208; Schischkin in Fl. SSSR, 3 (1935) 10; Fl. Kirgiz. 2 (1950) 236; Fl. Kazakhst. 2 (1958) 6; L. K. Dai in Fl. R. P. Sin. 11 (1961) 159; Fl. Tadzh. 2 (1963) 16. —*Cyperus serotinus* Rottb. Descr. pl. rar. (1772) 18; ej. Descr. and Icon. (1773) 21; Kük in Pflanzenreich [IV, 20], 101 (1936) 316; Kitag. Lin. Fl. Mansh. (1939) 116. —**Ic.**: Fl. Tadzh. 2, Plate II, figs. 1-3.

Described from India. Type in Copenhagen (?).

Along banks of rivers, marshy meadows and around ditches.

IA. Mongolia: ***Ordos*** (Huang He river valley, 1871—Przew.; Ulan-Morin river, 23 VIII 1884—Pot.; 75 km south of Dzhasak town, on bank of Lake Tautykhaitsi, 17 VIII 1957—Petr.).

IB. Kashgar: ***Nor.*** (Kurlya, 24 III 1929—Pop.; 4 km from Yuili [Chiglyk], in a ditch, No. 8571, 2 VIII; Yuili, Lake Lobulok, in water, No. 8571, 8 VIII 1958—Lee and Chu); ***South.*** (Khotan, oasis, 1230 m, 12 X 1885—Pot.); ***East.*** (Turfan, 11 X 1879—A. Reg.; Khami desert, Bugas region, near brook, 22 VIII 1895—Rob.).

IIA. Junggar: ***Jung. Alt.*** (Emel'-Toli, near water, No. 2987, 15 VIII 1957—Kuan); ***Tien Shan*** (Bayandai near Kul'dzha, 11 VII 1877; Dzhagastai, 600-900 m, 6 VIII 1877; Nilki, 30 VI 1879—A. Reg.; Arystan on Ili river, 7 VIII 1878—Fet.; 30 km south-east of Bole, along border of ditch, No. 4772, 29 VIII 1957—Kuan); ***Jung. Gobi*** (south.: 3-4 km east of St. Kuitun settlement, Khonkada area, marshy meadow, 7 VII 1957—Yun.; 3 km east of Kuitun, on a wet site, No. 427, 7 VII 1957—Kuan).

General distribution: Aral-Casp. (south.), Balkh. region, North Tien Shan (Chui valley); Europe (south.), Near East (Afghanistan), Caucasus, Mid. Asia, Far East (south.), China (Dunbei, North, North-west., Centr., South-west., Taiwan), Himalayas (Kashmir), Korea, Japan and Indo-Malay.

3. **Acorellus** Palla

in Koch, Syn. ed. 3, 3 (1907) 2557.

1. **A. pannonicus** (Jacq.) Palla in Koch, Syn. ed. 3, 3 (1907) 2557; Schischkin in Fl. SSSR, 3 (1935) 22; Fl. Kirgiz. 2 (1950) 240; Fl. Kazakhst. 2 (1958) 10. —*Cyperus pannonicus* Jacq., Fl. Austr. 5, App. (1778) 24; Krylov, Fl. Zap. Sib. 3 (1929) 378; Kük. in Pflanzenreich [IV, 20], 101 (1936) 319; Kitag. Lin. Fl. Mansh. (1939) 115. —*Juncellus pannonicus* (Jacq.) Clarke in Kew Bull. Add. ser. 8 (1908) 3; L. K. Dai in Fl. R. P. Sin. 11 (1961) 161. —**Ic.**: Jacq., l.c. Table 6; Fl. Kazakhst. 2, Plate 1, fig. 6.

Described from Hungary. Type in Vienna (?).

On wet and marshy sites; in depressions and in lower hill belt.

IA. Mongolia: ***Alash. Gobi*** (nor. Alashan, on sands and saline sites, 9 IX 1880—Przew.; Bayan-Khoto, Tengeri sands, Kheiyan'chi, 14 VIII 1958—Petr.); ***Khesi*** (Sachzhou oasis, along irrigation ditches and small swamps, 1110 m, 6 VIII 1895—Rob.).

IB. Kashgar: ***Nor.*** (Bai, 14, VIII 1929—Pop.; near Lake Tikenlik in Chiklyk, No. 8621, 10 VIII 1958—Lee and Chu); ***East.*** (alongside Lake Bagrashkul', in depression, No. 6181, 26 VII 1958—Lee and Chu).

IIA. Junggar: ***Jung. Gobi*** (west.: Junggar exit, near Kzyl-tuz meteorological station, solonchak-like grass plot, 16 VIII 1957—Yun.).

General distribution: Aral-Casp., Balkh. region, North and Centr. Tien Shan (Koisary); Europe, Caucasus, Mid. Asia (excluding Pamiroalay), West. Sib. (south, excluding Altay), China (Dunbei, North Centr.-north).

Note: In "Flora of China" (L. K. Dai, l.c.), this species has been cited for Inner Mongolia but no precise location indicated.

4. Pycreus Beauv.

Fl. Oware, 2 (1807) 48.

1. Spikelet not more than 2 mm broad; glumes almost continuous except for keel, dark purple, usually 1.5 mm long. Nut up to 0.8 mm long ....2. **P. nilagiricus** (Hochst. ex Steud.) Camus.
+ Spikelet about 3.5 mm broad; glume purple only along fringes, colourless between fringes and keel; 2.5-3 mm long. Nut about 1.4 mm long ....1. **P. korshinskyi** (Meinsh.) V. Krecz.

1. P. korshinskyi (Meinsh.) V. Krecz. in Bot. mater. Gerb. Bot. inst. AN SSSR, 7, 2 (1937) 27; Fl. Kirgiz. 2 (1950) 235; Fl. Kazakhst. 2 (1958) 5; Fl. Tadzh. 2 (1963). 15. —*P. sanguinolentus* f. *humilis* (Miq.) L. K. Dai and f. *rubro-marginatus* (Schrenk) L. K. Dai in Fl. R. P. Sin. 11 (1961) 171. —*P. sanguinolentus* auct. non Nees: Forbes and Hemsley, Index Fl. Sin. 3 (1905) 206, p.p. —*P. eragrostis* auct. non Palla: Schischkin in Fl. SSSR, 3 (1935) 7. —*Cyperus korshinskyi* Meinsh. in Acta Horti Petrop. 18, 1 (1901) 235. —*C. flavescens* var. *rubro-marginatus* Schrenk in Fisch. et Mey. Enum. pl. nov. 1 (1841) 1. —*C. rubro-marginatus* Meinsh. l.c. 235. —*C. sanguinolentus* f. *humilis* (Miq.) Kük. and f. *rubro-marginatus* (Schrenk) Kük. in Pflanzenreich [IV, 20], 101 (1936) 386. —**Ic.**: Fl. Kirgiz. 2, Plate 42.

Described from Junggar. Type in Leningrad.

On wet sites along banks of rivers and lakes; in the lower hill belt.

IA. Mongolia: ***Ordos*** (Dzhasygen-Qaidam area, 30 VIII 1884—Pot.; 10 km south-west of Ushin town, on edge of puddle, among thickets, 4 VIII; 50 km south of Dzhasak town, meadow with puddle among sand dunes, 17 VIII 1957—Petr.); ***Khesi*** (Sutcheou, 15 IX 1890—Martin).

IB. Kashgar: ***Nor.*** (Bai, 14 VIII 1919—Pop.); ***East.*** (south. fringe of Khami oasis, Bugas village, near brook, 480 m, 19 VIII 1895—Rob.; Lyamusin in Pichan, 480 m, No. 6658, 13 VI 1958—Lee and Chu).

IIA. Junggar: ***Tien Shan*** (Dzhagastai, 6 VIII 1877—A. Reg.); ***Jung. Gobi*** (south.: in Kuitun region, No. 3701, 12 X 1956—Ching); ***Dzhark.*** (Khoyur-Sumun, south of Kul'dzha, 26 V; Kul'dzha, 5 VII 1877—A. Reg.).

IIIA. Qinghai: ***Nanshan*** (Lovachen town, 26 VII 1909—Czet.).

General distribution: Balkh. region, Jung.-Tarb., North and Centr. Tien Shan; Caucasus, Mid. Asia, Far East (south.), China (entire), Korea and Japan.

Note: *P. korshinskyi* is very close to the South Asian species *P. sanguinolentus* (Vahl) Nees, the former differing only in capitate inflorescence and very short spikelets.

2. P. nilagiricus (Hochst. ex Steud.) E. G. Camus in Lecomte, Fl. Indo-Chine, 7, *1* (1912) 32; Schischkin in Fl. SSSR, 3 (1935) 6; Fl. Kirgiz. 1 (1950) 235; Fl. Kazakhst. 2 (1958) 5. —*P. globosus* var. *nilagiricus* (Hochst. ex Steud.) Clarke in Journ. Linn. Soc. (London) Bot. 21 (1884) 49; idem, in Forbes and Hemsley, Index Fl. Sin. 3 (1905) 204; L. K. Dai in Fl. R. P. Sin. 11 (1961) 167. —*Cyperus nilagiricus* Hochst. ex Steud. Syn. Pl. 2 (1855) 2; Kitag. Lin. Fl. Mansh. (1939) 115. —*C. globosus* var. *nilagiricus* (Hochst. ex Steud.) Kük. in Pflanzenreich [IV, 20], 101 (1936) 355. —*C. globosus* auct. non All.:

? Pampanini, Fl. Carac. (1930) 81; ? Chen and Chou, Rast. pokrov r. Sulekhe (Plant Cover of the Sulekhe River), (1957) 91.

Described from southern India (Nilgiri hills). Type in London (K). Isotype in Leningrad.

Along banks of rivers and lakes; in lower hill belt.

IA. Mongolia: ***Ordos*** (10 km south-west of Ushin town, on edge of puddle, 4 VIII 1957—Petr.).

IIA. Junggar: ***Tien Shan*** (Ili-Chapchal, on slightly saline soil, No. 3069, 5 VIII 1957—Kuan); ***Dzhark.*** (Kul'dzha, 300 m, 24 VIII 1878—A. Reg.); ***Balkh.-Alak.*** (around Dachen [Chuguchak], No. 2847, 10 VIII 1957—Kuan).

IIIA. Qinghai: ***Nanshan*** (Lovachen town, on swamp and silty soil, 26 VII 1908—Czet.).

General distribution: Balkh. region, Jung.-Tarb.; Mid. Asia (very rare), Far East (south.), China (Dunbei, North, North-west., Cent. East., South-west.), Himalayas (west.), Korea, Japan, Indo-Malay., Afr. (east.), Austral.

5. **Eriophorum** L.

Sp. pl. (1753) 52. Lo; idem, gen. Pl., ed. 5 (1754) 27.

1. Inflorescence with single spike ....2.
\+ Inflorescence with several spikes ....4.
2. Plant densely caespitose, without creeping rhizome, a few stems together; bristles ashy-white ....1. **E. brachyantherum** Trautv. et C. A. Mey.
\+ Plant loosely caespitose, with rather long creeping rhizome, stems usually single; bristles pure white ....3.
3. Glumes light grey, monochromatic; lower ones reflexed after flowering...2. **E. humile** Turcz. ex Steud.
\+ Glumes dark grey, nearly black, with narrow light-coloured margin; lower ones erect after flowering.....5. **E. scheuchzeri** Hoppe.
4(1) Peduncles smooth, flattened ....4. **E. polystachion** L.
\+ Peduncles scabrous, cylindrical or angular, sometimes slightly flattened ....3. **E. latifolium** Hoppe.

1. **E. brachyantherum** Trautv. et C. A. Mey. Fl. ochot. phaen. (1856) 98 Yuzepchuk in Fl. SSSR, 3 (1935) 34; Grubov, Consp. fl. MNR (1955) 79. —*E. callitrix* auct. non Cham.: Krylov, Fl. Zap. Sib. 3 (1929) 386, p.p. —**Ic.**: Fl. SSSR, 3, Plate II, fig. 7.

Described from Far East. Type in Leningrad (LE).

In marshy lands.

IA. Mongolia: ***Depr. Lakes*** (Ulangom's region, in a brook, 3 VII 1884—Pot.), Mong. Alt.

General distribution: Arct., Europe (north), East. Sib. (excluding Sayan mountains), Far East, North Mong., North Amer.

2. **E. humile** Turcz. ex Steud. Syn. pl. Cyper. (1855) 128; Yuzepchuk in Fl. SSSR, 3 (1935) 33; Grubov, Consp. fl. MNR (1955) 79. —*E. altaicum* auct. non Meinsh.: Krylov, Fl. Zap. Sib. 3 (1929) 386.

Described from East. Siberia (Sayan mountains). Type in Leningrad (LE).

In alpine marshy meadows.

IA. Mongolia. *Mong. Alt.* (basin of Bulugun river, Kharagaitu-Khutul' pass, alpine meadow and rubble placers with patches of snow, 24 VII 1947—Yun), Khobdo.

General distribution: Jung.-Tarb.; West. Sib. (Altay), East. Sib. (south.), Far East, North Mong. (Hent., Hang.).

3. **E. latifolium** Hoppe. Bot. Taschenb. (1800) 108; Danguy in Bull. Mus. nat. hist. natur. 20 (1914) 144; Forbes and Hemsley, Index Fl. Sin. 3 (1905) 256; Krylov, Fl. Zap. Sib. 3 (1929) 390; Yuzepchuk in Fl. SSSR, 3 (1935) 29; Kitag. Lin. Fl. Mansh. (1939) 117; Grubov, Consp. fl. MNR (1955) 79, p.p., excl. pl. alt. mong. and chobd.; Tang and Wang in Fl. R. P. Sin. 11 (1961) 37. —**Ic.**: Fl. SSSR, 3, Plate II, fig. 2.

Described from Europe. Type in Berlin (?).

In swampy meadows and on river banks.

IA. Mongolia: *Cis-Hing.* (Arshan town, in a wet meadow, No. 317, 14 VI 1951—Chang; "Vallée du Djatan-gol, bords de la riviere, 28 VI 1896—Chaff".—Danguy, l.c.); *East. Mong.* (near Khailar town, wet site, 4 VII 1950, 1951—Chang; "Inner Mongolia"—Tang and Wang, l.c.)

General distribution: Europe, Caucasus, ? West. Sib., East. Sib. (south.), Far East (south.), North Mong., China (Dunbei), Korean peninsula (north).

Note: The reference in "Flora of China" (Tang and Wang, l.c.) to this species growing in Inner Mongolia has been cited with no precise location.

4. **E. polystachion** L. Sp. pl. (1753) 52; ?Kitag. Lin. Fl. Mansh. (1939) 117. —*E. angustifolium* Honck. Verz. Gew. Deutschl. (1782) 153. —*E. angustifolium* Roth, Fl. Germ. 2 (1789) 63; Forbes and Hemsley, Index Fl. Sin. 3 (1905) 255; Krylov, Fl. Zap. Sib. 3 (1929) 388; Yuzepchuk in Fl. SSSR, 3 (1935) 29; Grubov, Consp. fl. MNR (1955) 79; Fl. Kazakhst. 2 (1958) 10. —*E. latifolium* auct. non Hoppe: Grubov, l.c. 79, p.p. —**Ic.**: Fl. SSSR, 3, Plate II, fig. 1.

Described from Europe. Type in London (Linn.).

In swamps and on marshy banks of rivers and lakes.

IA. Mongolia: *Bas. Lakes* (Ulangom's region, in marsh and in brook, 2 and 3 VII; Tszusylan, in marsh, 16 VII 1879—Pot.); *Mong. Alt.* (Dain-Gol lake, south-west, bank, 29 VII 1908—Sap.).

IIA. Junggar: *Alt. Region* (Kongeity river, 18 IX 1876—Pot.).

General distribution: Jung.-Tarb.; Arct., Europe, Caucasus, West. Sib., East. Sib., Far East, Nor. Mong. (Hent., Hang., Mong.-Daur.), China (Dunbei), Korean peninsula (north), North Amer.

5. **E. scheuchzeri** Hoppe. Bot. Taschenb. (1800) 104; Danguy in Bull. Mus. nat. hist. natur. 20 (1914) 144; Krylov, Fl. Zap. Sib. 3 (1929) 383; Yuzepchuk in Fl. SSSR, 3 (1935) 36; Fl. Kirgiz. 2 (1950) 243; Grubov, Consp. fl. MNR (1955) 79; Fl. Kazakhst. 2 (1958) 11, p.p. —**Ic.**: Fl. SSSR, 3, Plate II, fig. 11.

Described from Europe. Type in Berlin (?).

In marshy lands.

IA. Mongolia: *Khobd.* ("upper Sagliin-Gol"—Grubov, l.c.).

General distribution: Jung.-Tarb., Nor. and Cent. Tien Shan (extremely rare); Arct., Europe (north and mountains of Centr. Europe), West. Sib. (excluding Sayan mountains), Far East, North Amer., North Mong. (Fore Hubs, Hent.)

6. **Trichophorum** Pers.

Syn. 1 (1805) 70.

1. Stigmas three. Nuts trigonous..2. **T. pumilum** (Vahl) Schinz et Thell.
+ Stigmas two. Nuts biconvex.......1. **T. distigmaticum** (Kük.) Egor.

1. **T. distigmaticum** (Kük.) Egor. comb. nov. —*Scirpus pumilus* ssp. *distigmaticus* Kük. in Acta Hort. Gothoburg. 5 (1930) 34. —*S. distigmaticus* (Kük.) Tang and Wang in Fl. R. P. Sin. 11 (1961) 32.

Described from South-west. China (north Sichuan). Type in Göteberg (?).

Along river banks in high-altitude belt.

IIIA. Qinghai: ***Nanshan*** (on way from Alashan to Kukunor lake [1908]—Czet.).

IIIB. Tibet: ***Weitzan*** (hills along Bychu river, 18 VI; hills along Talachu river, 6 VI 1884, Przew.; basin of Yantszytozyan river, Yugin-do area, 3990 m, 13 V 1901—Lad.); ***South***. (Gyangtze, No. 62, VI-IX 1904—Walton).

General distribution: China (North-west., South-west.).

2. **T. pumilum** (Vahl) Schinz et Thell. in Vierteljahrsschr. Nat. Gesellsch. Zurich, 66 (1921) 265; Rozhev. in Fl. SSSR, 3 (1935) 38; Fl. Kirgiz. 2 (1950) 243; Grubov, Consp. fl. MNR (1955) 79; Fl. Kazakhst. 2 (1958) 12; Egorova in Bot. mater. Gerb. Bot. inst. AN SSSR, 19 (1959) 79; Fl. Tadzh. 2 (1963) 30; Ikonnikov, Opred. rast. Pamira (Key to Plants of Pamir) (1963) 75. —*Scirpus pumilus* Vahl, Enum. pl. 2 (1806) 243; Danguy in Bull. Mus. nat. hist. natur. 20 (1914) 144; Krylov, Fl. Zap. Sib. 3 (1929) 398; Tang and Wang in Fl. R. P. Sin. 11 (1961) 32. —*Scirpus pumilus* var. *distigmaticus* auct. non Kük.; ?Persson in Bot. Notis. (1938) 275. —Ic.: Fl. Tadzh. 2, Plate VI, figs. 1-3.

Described from Switzerland. Type in Copenhagen (?).

In wet, often saline meadows, near springs, along marshy river banks.

IA. Mongolia: ***Mong. Alt.*** (Bulugunsk region, meadow solonchaks on fringes of Tsukhul-Nor lake, 22 IX 1930—Bar.); Cent. Khalkha; ***East. Mong.*** (vallée du Kérulen, alt. 900 m, 3 VI 1896—Chaff."—Danguy, l.c.; "Inner Mongolia"—Tang and Wang, l.c.).

IB. Kashgar: ***West.*** (north of bridge on Sufu-Barkhetai highway, in Kashgar region, on floodplain, 1300 m, No. 00016, 22 IV 1959—Lee; upper course of Tiznafa river, Kyude village, 161 km from Kargalyk on Tibet highway, short-grass meadows along river, 1 VI 1959—Yun.; 27 km west of Maigaita, in Jarkand-Darya, 3900 m, No. 00482, 3 VI 1959—Lee).

IIA. Junggar: ***Tien Shan*** (20-25 km south of Urumchi, Ulumbai region, in swampy meadow fed by a spring 2 VI 1957—Yun.); ***Jung. Gobi*** (south.: midcourse of Taldy river, 2100 m, 26 V 1879—A. Reg.; east.: floodplain of Bodonchi river 2-3 km south of Bodonchin Khure, saline meadow, No. 9769, 1947—Yun.; Uienchi somon, Borotsonchzhi area, saline meadow, 13 IX 1948—Grub.); ***Dzhark.*** (Ili, bank, west of Kul'dzha, V 1877—A. Reg.; Kul'dzha, 1879—A. Reg.).

IIIA. Qinghai: ***Nanshan*** (Sharagol'dzhin river, in wet meadow, 2700 m, 13 VI 1894—Rob.); ***Amdo*** (upper course of Huang He river [Dzurge-Gol river], on sandy sites near springs 14 IV 1880—Przew.).

IIIB. Tibet: ***Chang Tang*** (upper course of Khotan-Darya river, 10-12 km west of Shakhidulla, short-grass meadow near river, 3900 m, 3 VI; Raskemdar'ya river valley, near Mazar settlement, saline meadow, 4 VI 1959—Yun.).

IIIC. Pamir (Kenkol river, Tom-Kara region, marshy hummocky bank of brook, 14 VI 1909—Divn.; "Pamir, Tashkorghan, Jurgal, about 3510 m, 29 VI 1935"—Persson, l.c.).

General distribution: Aral-Casp., Balkh. region, Jung.-Tarb., Cent. Tien Shan, East. Pam.; Europe (Centr. Europ. mountains, Urals), Near East, Caucasus (west.), Mid. Asia (West. Tien Shan, Pamiroalay), West. Sib. (south.), East. Sib. (south., excluding Sayan mountain range), North Mong. (Hang., Mong.-Daur.), China (North).

Note: In "Flora of China" (Tang and Wang, l.c.), this plant has been cited for Inner Mongolia without precise location, and also for West. Tibet.

7. **Scirpus** L.

Sp. pl. (1753) 47; idem, Gen. pl. ed. 5 (1754) 26.

1. Annual, with slender or filiform stem. Nuts longitudinally or transversely ribbed .... 2.
+ Perennial, with thick stem. Nuts smooth .... 3.
2. Bract leaf usually 5 mm long, less often up to 1 cm. Nuts 0.7-1 mm long, with fine oblong ribs and pearly lustre; perianth bristles absent .... 7. **S. setaceus** L.
+ Bract leaf up to 10 cm long. Nuts 1.2-1.4 mm long, with fine transverse ribs, without pearly lustre; perianth bristles present or sometimes absent .... 8. **S. supinus** L.
3. Bract leaves flat, broad. Inflorescence with many blackish-green spikelets .... 4.
+ Bract leaves trigonous or subulate. Spikelets brownish .... 5.
4. Spikelets ovate, rather obtuse .... 2. **S. sylvaticus** L.
+ Spikelets oblong-ovate, acuminate .... 1. **S. orientalis** Ohwi.
5. Stem flattened-trigonous, nearly plane, somewhat winged along ribs; sheath surrounding base of shoots transforming into long and flat blades. Bract leaf 15-20 cm long, many times longer than inflorescence .... 3. **S. ehrenbergii** Boeck.
+ Stem trigonous or cylindrical; sheath surrounding base of shoots without or with short blade. Bract leaf less long, not longer than thrice length of inflorescence .... 6.
6. Stem cylindrical. Glume ciliate along margin; sometimes with bristles or purple warts along keel .... 4. **S. hippolyti** V. Krecz.
+ Stem trigonous. Glume not ciliate or weakly so along margin, smooth .... 7.
7. Inflorescence spreading, with many spikes; spikelets usually single on stalks. Perianth bristles enlarged towards top where they are pectinate-hairy fimbriate .... 5. **S. litoralis** Schrad.
+ Inflorescence capitate or with rather elongated branches, with 2-5 aggregated spikelets; less often, spikelets single. Perianth bristles not fimbriate, retrorsely coenate .... 8.
8. Rhizome long, creeping. Stem near inflorescence 1-2.5 mm in diameter. Inflorescence with branches, sometimes capitate. Glume membranous, brownish throughout .... 9. **S. triqueter** L.

\+ Rhizome shortened, not creeping. Stem near inflorescence about 5 mm in diameter. Inflorescence invariably capitate. Glume coriaceous, pale green, brownish only along margin.........6. **S. mucronatus** L.

Subgenus **1. Scirpus**

1. **S. orientalis** Ohwi in Acta Phytotax. and Geobot. 1 (1932) 76; Kitag. Lin. Fl. Mansh. (1939) 123; Grubov, Consp. fl. MNR (1955) 79. —*S. sylvaticus* var. *maximowiczii* Rgl. Tent. Fl. Ussur. (1861) 161; Tang and Wang in Fl. R. P. Sin. 11 (1961) 9. —*S. radicans* auct. non Schkuhr: Rozhev. in Fl. SSSR, 3 (1935) 45, p.p., quoad var. *maximowiczii.* —**Ic.**: Fl. R. P. Sin. 11, Table II, 7-10.

Described from Japan. Type in Kyoto.

IA. Mongolia: ***East. Mong.*** (Khailar town, Nunlin'tun' village, near water, 10 VI 1951—Wang; "East. Mong."—Grubov, l.c.).

General distribution: Far East (south.), Nor. Mong., China (Dunbei, Nor., Nor.-west.), Korea and Japan.

Note: Species morphologically poorly differentiated from *S. sylvaticus* L.

2. **S. sylvaticus** L. Sp. pl. (1753) 61; Krylov, Fl. Zap. Sib. 3 (1929) 408; Rozhev. in Fl. SSSR, 3 (1935) 44; Fl. Kazakhst. 2 (1958) 15. —**Ic.**: Fl. SSSR, 3, Plate III, fig. 2.

Described from Europe. Type in London (Linn.).

In marshy lands.

IIA. Junggar: ***Balkh.-Alak.*** (around Dachen town, in marshy land, No. 2839, 10 VIII 1957—Kuan).

General distribution: Aral-Casp., Jung.-Tarb.; Europe, Asia Minor, Caucasus, West. Sib. (south.), East. Sib. (south.).

Subgenus **2. Isolepis** (R. Br.) Pax

3. **S. ehrenbergii** Boeck. in Linnaea, 36 (1870) 712; Krylov, Fl. Zap. Sib. 3 (1929) 406; Rozhev. in Fl. SSSR, 3 (1935) 51; Fl. Kazakhst. 2 (1958) 18; Tang and Wang in Fl. R. P. Sin. 11 (1961) 18. —**Ic.**: Fl. SSSR, 3, Plate IV, fig. 6; Fl. R.P. Sin. 11, Table VII.

Described from Europe (Transvolga). Type in Leningrad.

IB. Kashgar: ***East.*** (Yarkhu, in Turfan region, 100 m, No. 5567, 9 VI 1958—Lee and Chu).

General distribution: Balkh. region; Europe (Europ. part of USSR-Transvolga), West. Sib. (south.), China (North-west.).

4. **S. hippolyti** V. Krecz. in Bot. mater. Gerb. Bot. inst. AN SSSR, 7, *1* (1937) 28; Fl. Kazakhst. 2 (1958) 17. —? *S. validus* Vahl, Enum. pl. 2 (1805) 268; Tang and Wang in Fl. R. P. Sin. 11 (1961) 19. —*S. lacustris* auct. non L.: Henderson and Hume, Lahore to Yarkand (1873) 339; Forbes and Hemsley, Index Fl. Sin. 3 (1905) 250; Pampanini, Fl. Carac. (1930) 81;

Persson in Bot. Notis. (1938) 276; Chen and Chou, Rast. pokrov r. Sulekhe (Plant Cover of Sulekhe River) (1957) 91. —*S. tabernaemontanii* auct. non Gmel.: Krylov, Fl. Zap. Sib. 3 (1929) 405, p.p.; Fl. Kirgiz. 2 (1950) 247, p.p.; Rozhev. in Fl. SSSR, 3 (1935) 47, p.p.; Kitag. Lin. Fl. Mansh. (1939) 123; Grubov, Consp. fl. MNR (1955) 80, pro max. p.; Egorova in Bot. mater. Gerb. Bot. inst. AN SSSR, 19 (1959) 80. —*Schoenoplectus validus* (Vahl) Ovcz. et Czuk. in Fl. Tadzh. 2 (1963) 44. —**Ic.**: Fl. R. P. Sin 11, Table IX, 8-13 (sub *S. validus*).

Described from West. Kazakhstan. Type in Leningrad.

Along saline marshes and marshy meadows, depressions among solonchak, sometimes among puffed solonchak, along banks of water reservoirs and in water, along brooks and near springs; from lowlands to lower hill belt—1500 m above sea level (ascending to 2500 m above sea level in Qinghai and Qaidam).

IA. Mongolia: ***Khobd., Mong. Alt.*** (valley of Bulugun river, marshy land, 18 IX 1930—Bar.); ***Cis-Hing.*** (near Yaksha railway station, meadow lake, 19 VIII 1902—Litw.; Arshan town, in flowing water of hot springs, 15 VI 1950—Chang); ***Cent. Khalkha*** (cent. Kerulen, above Bars-Khoto, in meadows, 1899—Pal.; Mongolia, Ulkhuin-bulun area, in puddles, 12 VIII 1925—Gus.); ***East. Mong.*** (Buir-norsk trough, on wet meadow, 29 VII 1899—Pot. and Sold.; around Khailar town, near and in water, No. 586, 8 VI 1951—S. H. Li et al. (1951); Shilin-Khoto, 1959—Ivan.); ***Depr. Lakes*** (around Ubsa lake, Telin-Gol river, along banks, 9 VIII 1879—Pot.; Shargin-Gobi, near Gol-ikhe region, saline marsh, 5 IX 1930—Pob.; same site, Tongulak-Bulak spring, on road, among puffed solonchak, 5 IX 1948—Grub.); ***Gobi-Alt.*** (Baintukhum region, depressions among solonchaks, 4 VIII 1931—Ik.-Gal.; 35-40 km west-nor.-west of Dalan-Dzadagad, saline meadows, No. 10224, VII 1943—Yun.); ***East. Gobi*** (Undur-shili somon, Toli-Bulak region, saline meadow near spring, 27 VII 1946—Yun.); ***West. Gobi*** (Bilekhu-Bulak region 30-35 km east of Tsagan-Bogdo town, saline meadow near spring, 31 VII; same site, Khatun-Suudal well, edge of spring in barren land, 6 VIII; Ikhe-Tszaram region, swampy saline meadows, 19 VIII 1943—Yun.); ***Alash. Gobi*** (Edzin-Gol river between Mumin and Tufyn, 11 VII 1886—Pot.; Dyn-yuan'-in [Bayan-Khoto] oasis, along a brook, 3 VI 1908—Czet.; Bayan-Khoto, Tengeri sands, 7 VIII 1958—Petr.); ***Ordos*** (valley of Huang He river, 11 VIII 1871—Przew.; 10 km south-east of Inchuan', in standing water, 18 VII; 10 km south-west of Ushin town, along edge of puddle, 4 VIII; 75 km south of Dzhasak town, on bank of Tautykhaitszy lake, 17 VIII 1957—Petr.); ***Khesi*** (Kheikho river below Gaotai, 20 VI; Khechen village, flooded meadows along Edzin-Gol river, 27 VI 1886—Pot.; "Sulekhe river"—Chen and Chou, l.c.).

IB. Kashgar: ***North*** (Uchturfan, in a garden, 14 V 1908—Divn.; Pichan district, 100 km west of Khando, near a pond, 460 m, No. 6666, 13 VI; 10 km south-west of Kurl', near a canal, 950 m, No. 5897, 17 VII; Bai district, Koe village, near water, No. 8141, 1 IX; 5 km nor. of Vensu in Aksu district, 1000 m, No. 8972, 30 IX 1958—Lee and Chu); West. (Yangishar, 25 VII 1927—Pop.; from Kashgar to Faizabad, on wet sites, No. 7546, 27 IX 1958—Lee and Chu); ***South*** (Niya oasis, 1260 m, 4 VII 1885—Przew.; 10 km nor. of Kerii town, Bostan area, in a sasa, 15 V; 8 km west of Kerii town, on road to Chira, marshy valley floor, 16 V 1959—Yun.; Kerii-Chira road, in marshy land, No. 00112, 16 V 1959—Lee et al.; "Common near Yarkand city"—Henderson and Hume, l.c.; "Kashgar, about 25 km south of the town, about 1330 m, 1935"—Persson, l.c.); ***East.*** (Khami, in water, 29 V 1877—Pot.; Ledun in Khami region, near water, No. 459, 21 VI 1957—Kuan; 6 km south-east of "Khugan" state farm in Karashar, in water, No. 6085, 25 VII; near Bagrashkul' lake in a depression, No. 6176, 26 VII—Lee and Chu); ***Takla-Makan*** (6 km from Cherchen, 1200 m, No. 9509, 12 VI; 139 km from Ander in Niya district, in floodplain, 1393 m, No. 9587, 1959—Lee et al.).

IC. Qaidam: ***mount.*** (east. Qaidam, Tuguryuk region, 2580 m, 31 VII 1901—Lad.).

IIA. Junggar: ***Altay region*** (in a depression in Koktogoi region, No. 1864, 13 VIII; Altay [Shara-sume], Nos. 2318 and 2358, 20 and 25 VIII 1956—Ching); ***Jung. Alt.*** (Dzhair mountain range, Ker hills, 8 VII 1953—Mois.; nor.-west. margin of Dzhair mountain range 24 km nor.-east of Toli settlement, near a brook, in water, 5 VIII 1957—Yun.); ***Tien Shan*** (Piluchi, 900 to 1500 m, 24 IV; sources of Taldy river, 900-1200 m, 14 V 1879—A. Reg.; Urumchi-Ulapo, near water, No. 599, 2 VI; Kalmankure on Tekes river, No. 3333, 11 VIII 1957—Kuan); ***Jung. Gobi*** (south.: 2-3 km nor.-east of St. Kuitun settlement, on Shikho-Manas road, marshy meadows fed by spring, 30 VI 1957—Yun.; same site, in wet areas, No. 1123, 29 VI; in Kuitun region, in meadow, No. 272, 29 VI; 2 km north of Kuitun, along canal edge, No. 401, 6 VII; on bank of Manas river, No. 790, 11 VI; Shamyntsza, in a swamp, No. 953, 17 VI; Savan district, in Leyagukhu water reservoir, No. 1572, 25 VI 1957—Kuan); ***Zaisan*** (left bank of Ch. Irtysh river against Cherektas hills, on swamp, 11 VI 1914—Schischk.); ***Dzhark.*** (Kul'dzha, 1870; same site, VIII 1876; same site, 5 V 1877—A. Reg.; north of Kul'dzha, 3 VI 1877—A. Reg.).

IIIA. Qinghai: *Amdo* (upper course of Huang He river, in a marshy zone, 2400 m, 2 VI 1880—Przew.).

IIIB. Tibet: ***South.*** (Lhasa, 1904—H. F. Walton).

General distribution: Aral-Casp., Balkh. region, Jung.-Tarb., Nor. and Cent. Tien Shan; Europe (south.), Near East, Mid. Asia, Caucasus, West. Sib. (south., ?Altay), East. Sib. (south., ?Sayan mountain range), Far East (south.), Nor. Mong. (Hent., Hang., Mong.-Daur.), China (Dunbei, North, North-west., Centr., East., South-west.), Himalayas (west., Kashmir), ?North Amer.

Note: Plant widely distributed in Central Asia. Separation of this species from *S. lacustris* L. and *S. tabernaemontanii* Gmel., in our view is quite evident (see Kreczetowicz, l.c.). In general, however, the taxonomic position of this plant is variously interpreted by different authors. For example, the authors of "Flora of China" (Tang and Wang, l.c.) identify this rush with *S. validus* Vahl described from North America. This view is supported by P. N. Ovczinnikov and N. P. Chukavina in "Flora of Tadzhikistan". Koyama [Canad. Journ. Bot. 40 (1962) 927] treats it as *S. lacustris* ssp. *validus* (Vahl) T. Koyama var. *luxurians* (Miquel) T. Koyama, while Raymond [Biol. Skr. Dan. Vid. Selsk. 14, 4 (1965) 13] regards it as *S. lacustris* var. *luxurians* (Miquel) Raymond.

V. I. Kreczetowicz identified (in the herbarium) the plants covered here among *S. hippolyti* as *S. ciliatus* Steud.

Within Central Asia, *S. hippolyti* varies in dimensions of the inflorescence and number of spikelets (from a few to several) as well as in the nature of glume surface, which is either entirely smooth (in most specimens) or with some bristles or purple warts set along the midrib and, less often, scattered over the surface.

5. **S. litoralis** Schrad. Fl. Cerm. 1 (1806) 142; Ostenf. in Hedin, S. Tibet (1922) 90; Rozhev. in Fl. SSSR, 3 (1935) 52; Persson in Bot. Notis. (1938) 276; Fl. Kazakhst. 2 (1958) 18. — ?*S. subulatus* Vahl, Enum. pl. 2 (1805) 268; Tang and Wang in Fl. R. P. Sin. 11 (1961) 17. —*Schoenoplectus litoralis* Palla in Englers Bot. Jahrb. 10 (1889) 299; Fl. Kirgiz. 1 (1950) 248; Fl. Tadzh. 2 (1963) 46. —Ic.: Fl. Tadzh. 2, Plate IX, figs. 1-3.

Described from Europe. Type in Munich.

Along banks of rivers and canals and in marshy lands; on foothills, in lower mountain belt (up to 1300 m above sea level) and on plains.

IB. Kashgar: ***Nor.*** (Yarkand-Darya, Lai-mai floodplains, on wet clay, 1050 m, 15 VI; Yarkand-Darya, canals, 900 m, 21 VI 1889—Rob.; "Maralbashi, in a swamp about 1070 m, 14 VII 1932"—Persson, l.c.); ***South.*** (Niya oasis, 1260 m, 3 V; Keriya [15 VI] 1885—Przew.; 25 km nor.-west of Khotan, along road-to Karasai, strip of sasa, 25 V 1959—Yun.); ***East.*** (in Turfan region, near water, 120 m, No. 6620, 7 VI 1958—Lee and Chu); ***Takla Makan*** (Karakash district, in Karasai region, in swamp, No. 00148, 25 V 1959—Lee et al.); *Lob.-Nor* ("Lopnor, Kara-koshun, beneath Just-Chapghan, 816 m, 24 VI 1900—Hedin"—Ostenfeld, l.c.).

IIA. Junggar. ***Jung. Gobi*** (18-20 km south-east of Darabuta along Altay-Karamai road, slowing down streams of lower courses of Manas river, 20 VI 1957—Yun.).

General distribution: Aral-Casp., Balkh. region; Europe (south of West. Europe), ?Asia Minor, Near East, Caucasus, Mid. Asia, China (North, North-west., South-west.), Himalayas (Kashmir), Indo-Malay., Afr. (North), Austral.

6. **S. mucronatus** L. Sp. pl. (1753) 50; Rozhev. in Fl. SSSR, 3 (1935) 51; Persson in Bot. Notis. (1938) 276; Fl. Kazakhst. 2 (1958) 19; —*Schoenoplectus mucronatus* (L.) Palla in Englers Bot. Jahrb. 10 (1889) 299; Fl. Kirgiz. 2 (1950) 248; Fl. Tadzh. 2 (1963) 37. —**Ic.**: Fl. SSSR, 3, Plate IV, fig. 7.

Along river banks and in marshy lands.

IB. Kashgar: ***West.*** ("Kashgar, near river, about 1330 m, 18 VI 1933"—Persson, l.c.; "Opal, in swamp, about 1400 m, 15 VIII 1934"—Persson, l.c.).

General distribution: Europe (West. Europe), Near East (Iran), Caucasus, Mid. Asia (plains west of Pamiroalay, West. Tien Shan), Far East (south.), Korea, Japan, ?Austral.

Note: The reference to the occurrence of *S. mucronatus* in China cited in "Flora of the USSR" pertains to a closely related species, *S. triangulatus* Roxb.

7. **S. setaceus** L. Sp. pl. (1753) 49; Forbes and Hemsley, Index Fl. Sin. 3 (1905) 253; Krylov, Fl. Zap. Sib. 3 (1929) 400; Pampanini, Fl. Carac. (1930) 81; Rozhev. in Fl. SSSR, 3 (1935) 46; Fl. Kazakhst. 2 (1958) 16; Tang and Wang in Fl. R. P. Sin. 11 (1961) 29. —*Schoenoplectus setaceus* (L.) Palla in Koch, Syn. ed. 3, 3 (1907) 2538; Fl. Kirgiz. 1 (1950) 244; Fl. Tadzh. 2 (1963) 37. —**Ic.**: Fl. Tadzh. 2, Plate VII, figs. 5-7.

Described from Europe. Type in London (Linn.).

Along river banks.

IIA. Junggar: ***Altay region*** (Qinhe, No. 1669, 11 VIII 1956—Ching).

IIIA. Qinghai: ***Nanshan*** (Pinfan town, along river tributaries, 21, VII 1909—Czet.; "Qinghai"—Tang and Wang, l.c.).

General distribution: Aral-Casp., Balkh. region, Jung.-Tarb.; Europe (predominantly in south. half), Near East, Caucasus, Mid. Asia (West. Tien Shan, Pamiroalay), West. Sib. (south., ?Altay), China (North-west., South-west.), Himalayas.

8. **S. supinus** L. Sp. pl. (1753) 49; Danguy in Bull. Mus. nat. hist. natur. 20 (1914) 144; Krylov, Fl. Zap. Sib. 3 (1929) 401; Rozhev. in Fl. SSSR, 3 (1935) 53; Fl. Kazakhst. 2 (1958) 20; Tang and Wang in Fl. R. P. Sin. 11 (1961) 27, p.p., quoad var. *supinus*. —**Ic.**: Fl. SSSR, 3, Plate IV, fig. 11.

Described from Europe. Type in London (Linn.).

Along silty river banks and sinkholes in meadows.

IA. Mongolia: ***Depr. Lakes***

IIA. Junggar: ***Jung. Gobi*** (north: Bulun-Tokhoi, 20 VIII; Urungu river, 20 VIII; Ch. Irtysh, 26 VIII 1876—Pot.; left bank of Urungu river 85 km beyond Din'syan on road to Ertai, edge of meadow, 13 VII 1959—Yun.; "Bords de l'Irtich, 29 VIII 1896, Chaff."—Danguy, l.c.).

General distribution: Aral-Casp., Balkh. region; Europe, Asia Minor, ?Near East, Caucasus (Transcaucasus), West. Sib. (south., excluding Altay), East. Sib. (south-west., except for Sayan mountain range), Indo-Malay., North Amer.

9. **S. triqueter** L. Mant. I (1767) 29; Forbes and Hemsley, Index Fl. Sin. 3 (1905) 255; Rozhev. in Fl. SSSR, 3 (1935) 48; Kitag. Lin. Fl. Mansh. (1939) 123; Fl. Kazakhst. 2 (1958) 19; Tang and Wang in Fl. R. P. Sin. 11 (1961) 18. —*Schoenoplectus triqueter* (L.) Palla in Englers Bot. Jahrb. 10 (1889) 299; Fl. Kirgiz. 2 (1958) 247; Fl. Tadzh. 2 (1963) 41. —Ic.: Fl. SSSR, 3, Plate IV, fig. 5.

Described from Europe. Type in London (Linn.).

Along banks of lakes and in marshy habitats; on plains and on foothills.

IA. Mongolia: ***East. Mong.*** (Shilinkhoto town, steppe, 1957—Ivan.); ***Ordos*** (10 km southwest of Ushin town, on lake shore, 4 VIII; 50 km from Dzhasak town, meadow with puddle, 17 VIII 1957—Petr.).

IB. Kashgar: ***North*** (Aksu, 10 VIII 1929—Pop.); ***West.*** (Yangishar, 25 VIII 1929—Pop.); ***East.*** (nor.-east of Toksun, swamp, 300 m, No. 7338, 19 VI 1958—Lee and Chu).

General distribution: Aral-Casp., Balkh. region; Europe (south.), Near East, Caucasus, Mid. Asia (less often), Far East (south.), China (North, North-west, South, Hainan), Korea, Japan, Himalayas (West, Kashmir), ?North Amer.

8. Bolboschoenus (Aschers.) Palla

in Koch, Synops. ed. 3, 3 (1907) 2531.

1. Inflorescence umbellate; spikelets aggregated, generally on long peduncles, some on shortened, almost undeveloped peduncles. Stigmas three; nut convex-trigonous......1. **B. maritimus** (L.) Palla.
+ Inflorescence capitate, spikelets sessile; less often, inflorescence compound, umbellate, but then stigmas two and nut not trigonous.... 2.
2. Rhizome with large globose tubers. Glumes rusty-brown. Nut centrally impressed on both sides, 2.8-3.3 mm long......................... ....2. **B. planiculmis** (Fr. Schmidt) Egor.
+ Rhizome with or without tiny tubercles. Glumes yellowish-white. Nut planoconvex, 2.3-2.5 mm long....................3. **B. popovii** Egor.

1. **B. maritimus** (L.) Palla in Koch, Synops. ed. 3, 3 (1907) 2532; Rozhev. in Fl. SSSR, 3 (1935) 56; Fl. Kirgiz. 2 (1950) 251; Fl. Kazakhst. 2 (1958) 21; Fl. Tadzh. 2 (1963) 48. —*Scirpus maritimus* L. Sp. pl. (1753) 50; Krylov, Fl. Zap. Sib. 3 (1929) 406, p.p.; Persson in Bot. Notis. (1938) 276. —Ic.: Fl. SSSR, 3, Plate III, fig. 10.

Described from Europe. Type in London (Linn.). Plate VIII, fig. 17.

Along banks of water reservoirs, marshy lands, sometimes in saline soil; on plains, foothills and lower montane belt up to 1350 m above sea level.

IA. Mongolia: ***Mong. Alt.***

IB. Kashgar: ***North*** (Yarkand-Darya, among reeds. on sedimentary soil, 15 VI 1889—Rob.; Maralbashi, Chuderlik village, 6 V 1909—Divn.; Kurlya, 165 m, No. 6840, 20 VII; Shakh'yar district, north of Staryi Gimyn, near water, No. 8763, 17 IX 1958—Lee and Chu); West. (Yangishar, 9 VIII 1913—Knorring; same site, 25 VII 1929—Pop.; "Kashgar, in rice fields, 5 VIII 1925; Kashgar, riverside, about 1330 m, 18 VI 1933"—Persson, l.c.); ***East.*** (in "Krasnaya Zvezda" state farm in Turfan, on well-developed solonchak, 130 m, No. 5472, 30 V; Turfan, in water, No. 5477, 30 V; 10 km nor.-east of Toksun, saline meadow, Nos. 7258 and 7319, 15 and 16 VI 1958—Lee and Chu; Khami, 3-4 V to 22 VI 1879—?Pot.).

IIA. Junggar: ***Altay region*** (Shara-Sume, 800 m, No. 2695, 6 IX; south of Shara-Sume, 550 m, No. 2877, 8 IX 1956—Ching; Koktogoi, along spring, solonchak, 900 m, No. 10431, 10 VI; 1-2 km north of confluence of Kran and Irtysh rivers, No. 10620, 11 VII 1959—Lee et al.); ***Tien Shan*** (Piluchi, 900-1500 m, 24 IV; Kash river, between Ulutai and Nilka, 900-1200 m, 30 VI 1879—A. Reg.; Khaidyk-Gol river, Chubogorin-Nor area, swamp, 20 VIII 1893—Rob.; Ili-Chapchal, on somewhat saline soil, No. 3058, 5 VIII 1957—Kuan); ***Jung. Gobi*** (nor.: Ch. Irtysh, 26 VIII 1876—Pot.; valley of Urungu river, 3-4 km south of Bulun-Tokhoi settlement, along stream bank, 9 VII; same site, 85 km beyond Din'syan, floodplain, 13 VII 1959—Yun.; west.: Karamai-Shikhetsza, No. 3376, 24 VII 1956—Ching; 30 km north-north-east of Karamai, Dzheirak-Bulak area, solonchaks, 19 VI; 8-10 km of Darbata river on Altay to Karamai highway, near spring and in water, 20 VI; valley of Darbata river near its crossing with Karamai-Altay highway, solonchak reed thicket, 20 VI 1957—Yun.; south.: Savan dist., 15 and 24 km south of Tien Shan-Laoba railway station, on sand dunes, Nos. 197 and 982, 7 and 21 VI; same site, Seyadi, No. 1502, 16 VI; Savan district, No. 1577, 26 VI 1957—Kuan; east.: Nom, 2 VI 1877—Pot.); ***Zaisan*** (right bank of Ch. Irtysh below Burchum river, 15 VI 1914—Schischk.); ***Dzhark.*** (Kul'dzha, 1876—Golike; Shimpanzi village west of Kul'dzha, 8 V; sultan's garden near Kul'dzha, 12 VI 1877—A. Reg.).

General distribution: Aral-Casp., Balkh. region, Jung.-Tarb., Nor. and Cent. Tien Shan; Europe, Asia Minor, Near East, Caucasus, Mid. Asia, West. Sib. (south.), East. Sib. (south.), Far East, North Amer., Afr. (north), ?Austral.

2. **B. planiculmis** (Fr. Schmidt) Egor. comb. nov. —*Scirpus planiculmis* Fr. Schmidt in Mém. Ac. Sci. St. Petersb. 7, 12 (1868) 190; Rozhev. l.c. 48; Tang and Wang in Fl. R. P. Sin. 11 (1961) 7. —*B. compactus* Drob. in Tr. Bot. muzeya AN, 11 (1913) 92, p.p., excl. pl. europ.; Rozhev. in Fl. SSSR, 3 (1935) 57, p.p.; Fl. Kirgiz. 2 (1950) 251, p.p.; Grubov, Consp. fl. MNR (1955) 80; Fl. Kazakhst. 2 (1958) 21, p.p.; Egorova in Bot. mater. Gerb. Bot. inst. AN SSSR, 19 (1959) 80. —*S. maritimus* var. *affinis* auct. non Clarke: Clarke in Forbes and Hemsley, Index Fl. Sin. 3 (1905) 251. —*S. maritimus* var. *compactus* auct. non G. Mey.; Krylov, F. Zap. Sib. 3 (1929) 407. —*S. compactus* auct. non Hoffm.: Kitag. Lin. Fl. Mansh. (1939) 122.

Described from Sakhalin. Type in Leningrad. Plate VIII, fig. 15.

Along wet and marshy, often saline meadows, swamps, near springs, along banks of water reservoirs and in water; on plains, on foothills and in lower montane belt up to 1100 m above sea level; up to 2400 m above sea level in Amdo.

IA. Mongolia: ***Cis-Hing.*** (near Yaksha railway station, marsh, 19 VIII 1902—Litv.); ***East. Mong., Cent. Khalkha, Val. lakes*** (south. bank of Orok-Nur lake, 1 IX 1886—Pot.; Orok-Nor, border of lagoon, 1110 m, No. 303, 1925—Chaney; same site, wet lagoons near lake and solonetz, 4 VIII 1926—Tug.; west. border of Orok-Nur lake, saline meadow, 13 IX 1943—Yun.; Tsagan lake, on wet solonchak, 28 VI 1924—Pavl.); ***East. Gobi*** (20 km north of Hubsugul somon on road to Sain-Shandu, Baingiin-Bulak region, saline meadow, 28 VI 1941; Undur-Shili somon, Toli-Bulak region, on edge of a spring, 27 VII 1946—Yun.; Bailinmyao town, solonchak lowlands, 1959—Ivan.); ***West. Gobi*** (Trans-Altay Gobi, Khatun-Suudal collective, near a spring, 6 VIII 1943—Yun.); ***Alash. Gobi*** (Dyn'-yuan'-in [Bayan-Khoto] oasis, along banks, on silty soil, 29 V 1908—Czet.; Chzhunvei to Inchuan' road, right bank of Huang He river, 3 VII 1957—Kabanov; 10 km south-east of Inchuan', in water, 18 VII; 45 km south of Inchuan' town, saline meadow, 3 VII 1957—Petr.); ***Ordos*** (val. of Huang He river, 8 VIII 1871—Przew.; 10 km south-west of Ushin town, on edge of puddle, 4 VIII; 50 km south of

Dzhasak town, meadow with puddle among sand dunes, 17 VIII 1957—Petr.); ***Khesi*** (between Shai village and Fuiitin town, 5 VI; between Fuiitin and Gaotai, 6 and 17 VI; Edzin river beyond Gaotai, 20 VI; between Yant settlement and Huang He river, 25 VI 1886—Pot.).

IB. Kashgar: ***North*** (Uchturfan, 14 V 1908—Divn.); ***East.*** (in second region of Turfan dist., near a ditch, No. 5442, 26 V; in Turfan, Putougou settlement, near water, 100 m, No. 5510, 1 VI; Shizinko in Turfan, near water, No. 6624, 8 VI; Yarkhu, in Turfan, No. 5566, 9 VI; Khoraz in Toksun, near a pond, No. 7227, 9 VI; south of Dunkh, in Pichan, near water, No. 6680, 14 VI; Karashar, in region of Bagrashkul' lake, in water, 1100 m, No. 6086, 26 VII 1958—Lee and Chu).

IIA. Junggar: ***Altay region*** (Koktogoi-Ukagou, Nos. 1696 and 2253, 11 and 20 VIII 1956—Ching; Koktogoi, near water, 950 m, Nos. 10404 and 10405, 7 VI 1959—Lee et al.); ***Jung. Alt.*** (nor.-west. border of Dzhair mountain range, 24 km nor.-east of Toli settlement, near brook, in water, 5 VIII 1957—Yun.); ***Jung. Gobi*** (south: Savan district, Shikhetsz, No. 3768, 7 X 1956—Ching; Urumchi, in and near water, No. 579, 7 VI, bank of Manas river, on marshy sites, No. 791, 11 VI; Savan district, Shamyntsz, No. 950, 17 VI; 20 km nor. of Kuitun, No. 381, 6 VII 1957—Kuan; near Tsitai town, in swamp, No. 4453, 23 IX; 3-4 km nor. of St. Kuitun settlement, sasa zone, herb-sedge marshy meadow, 6 VII 1957—Yun.); ***Dzhark.*** (sultan's garden near Kul'dzha, 12 VI; Kul'dzha, 15 VI 1877; Suidun, 7 V 1878; Piluchi, 28 IV 1879—A. Reg.); ***Balkh.-Alak.*** (south of Dachen, No. 2857, 11 VIII; Dachen-Emel', No. 2860, 11 VIII 1957—Kuan).

IIIA. Qinghai: ***Amdo*** (upper course of Huang He river, in swamp, 2 VI 1880—Przew.).

General distribution: Aral-Casp., Balkh. region, Jung.-Tarb., Nor. and Cent. Tien Shan (very rare); Europe (south Europ. USSR), ?Near East, Mid. Asia (West. Tien Shan), West. Sib. (south.), East. Sib. (south.), Far East (south.), China (Dunbei, North, North-west., East., South-west.), Korea and Japan.

Note: Several authors have regarded this plant as *B. compactus* (Hoffm.) Drob. described from Western Europe. But then, *B. compactus* with compact inflorescence without rays is only a variant of *B. maritimus [B. maritimus* f. *compactus* (Hoffm.) Egor.]. The distribution ranges of the two forms coincide and, quite often, they are found within the same population.

B. planiculmis is well distinguished from *B. compactus* in styles with two stigmas and biconvex nuts which are impressed anteriorly and often posteriorly too (stigmas three and nuts trigonous in *B. compactus* as also in *B. maritimus*).

In Mongolia (Alashan, Ordos and Khesi), specimens are found of *B. planiculmis* which have an umbellate inflorescence, as in *B. maritimus.*

3. **B. popovii** Egor. sp. nov. —*B. affinis* (Roth) Drob. in Tr. Bot. muz. AN 16 (1916) 139, quoad pl. Asiae med.; Rozhev. in Fl. SSSR, 3 (1935) 57; Fl. Kirgiz. 2 (1950) 252; Grubov, Consp. fl. MNR (1955) 80; Fl. Kazakhst. 2 (1958) 22. —*B. strobilinus* (Roxb.) V. Krecz. in Fl. Tadzh. 2 (1963) 47, quoad pl. Asiae med. et centr. —*Scirpus affinis* auct. non Roth: Ostenfeld in Hedin, S. Tibet (1922) 90; Persson in Bot. Notis, (1938) 276. —*S. maritimus* auct. non L.: Walker in Contribs. U.S. Nat. Herb. 28 (1941) 600; Chen and Chou, Rast. pokrov r. Sulekhe (Plant Cover of Sulekhe River) (1957) 91. —*S. strobilinus* auct. non Roxb.; Tang and Wang in Fl. R. P. Sin. 11 (1961) 8. —Ic.: Fl. Tadzh. 2, Plate X, figs. 5-7. —Planta perennis, 20-60 (80) cm alt., rhizomate repenti. Caules triquetri, leves vel superne scabri, foliati. Folia plana, 2-8 mm lt. Inflorescentia e spiculus 1-20 confertis in capitulum constans. Spiculae (1) 1.3-1.5 (1.8) cm lg., oblongo-ovatae, acuminatae. Squamae late-ovatae, flaveolo-albae, non purpureo-striatae

vel vix purpureo-striatae, aristatae. Fructus plano-convexus late-ovatus vel rotundus, 2.3-2.5 mm lg.

Typus: China, Kaschgaria, Bugur, No. 795, 20 VIII 1929, M. G. Popov. In Herb. Inst. Bot. Acad. Sci. URSS (Leningrad) conservatur.

Affinitas. A specie proxima *S. strobilino* Roxb. (*S. affinis* Roth) differt: spiculis minoribus (1) 1.3-1.5 (1.8) cm lg. (nec 1.8-2.3 cm lg.) et squamis non purpureo-striatis vel vix purpureo-striatis (nec copiose purpureo-striatis).

Plate VIII, fig. 16.

On wet, quite often saline sites, near springs, along banks of water reservoirs and in water, on marshy lands; on plains, in foothills and in lower montane belt up to 1200 m above sea level.

IA. Mongolia: ***Depr. Lakes*** (valley of Dzergin river, solonchak, 18 VIII 1930—Bar.); ***Val. lakes, Gobi-Alt.*** (Bain-Tukhum area, depressions among solonchaks, 4 and 31 VIII 1931—Ik.-Gal.); ***West. Gobi*** (30-35 km from Tsagan-Bogdo town, Bilgekhu-Bulak area, saline meadow near a spring, 31 VII 1943—Yun.); ***Alash. Gobi*** ([camp site near Chirgu-Bulyk] well, 10 VIII 1873—Przew.; Koko-buryuk—Gantsy-obo, 20 VII; Kharasukhai area, 21 VII 1886—Pot.; Dyn-yuuan'-in [Bayan-Khoto] oasis, 1 VI 1908—Czet.; same site, Mintsin town, 15 VII; same site, Tengeri sands, 23 VII 1958—Petr.; Edzin-Gol river, Chzhargalante area, 17 VI 1909—Czet.; "Chung Wei [Chzhunvei town], on edge of a stream, No. 215, 1923—R. C. Ching"—Walker, l.c.); ***Khesi*** (Yan'chi village, near Yan'chi lake, 29 VI 1886—Pot.; "Sulekhe river"—Chen and Chou, l.c.).

IB. Kashgar: ***North*** (Yarkand-Darya, Lai-mai floodplains, on sand, 1050 m, 14 VI 1889—Rob.; Yuili, in Lobulok lake, No. 8569, 8 VIII; nor.-west of Sankuduk in Marbat [Maralbashi], dry river bed, No. 7783, 2 IX 1958—Lee and Chu; Bugur, 20 VIII 1929—Pop., typus!; Pichan trough, Chon-Karadzhal settlement, border of solonchak trough, grass plot near spring, 22 VI 1959—Yun.; Kashgar to Faizabad, No. 7547, 27 IX; 3 km nor.-west of Tsaokhu in Bugure, No. 8689, 1 IX 1958—Lee and Chu; "Maralbashi, 1150 m, 14 VI 1932"—Persson, l.c.); ***West.*** (Yandoma, 24 VII 1929—Pop.; Artush-Khalatsi, in a ditch, No. 09779, 22 VI 1959—Lee et al.); ***South.*** (Cherchen district in Tatrak region, in floodplain, 1213 m, No. 9096, 2 VI 1959—Lee et al.); ***Takla Makan*** (6 km nor.-west of Cherchen district, No. 9508, 12 VI 1959—Lee et al.); ***Lobnor*** (islet in Lobnor lake, near water, 380 m, No. 10125, 23 IX 1959—Lee et al.; "Lob-nor, Kara-Koshun, 24 VI 1900, S. Hedin"—Ostenfeld, l.c.).

IIA. Junggar: ***Altay region*** (south of Shara-Sume, 700 m, No. 2767, 6 IX 1956—Ching); ***Jung. Gobi*** (south.: Ula-usu, saline swamp, 319 m, 2 VII 1957—Kuan); ***Dzhark.*** (near Kul'dzha, 1880—A. Reg.).

General distribution: Aral-Casp., Balkh. region, Jung.-Tarb.; Europe (Europ part of USSR—Transvolga), Near East, Caucasus (Sara island), Mid. Asia.

'**Note:** Morphologically (see above) and in distributional range, the species under description differs distinctly from the plant described from eastern India (Ganga river) as *Scirpus strobilinus* Roxb. [= *Bolboschoenus strobilinus* (Roxb.) V. Krecz.]. We have not studied the type of *S. strobilinus*. The Herbarium of the Botanical Institute, Academy of Sciences, USSR (Leningrad), however, contains specimens from this region corresponding to the description of *S. strobilinus* and most probably belonging to it.

Bolboschoenus affinis (Roth) Drob. (quoad typus) (= *Scirpus affinis* Roth) can be regarded as a synonym for *B. strobilinus*

9. **Blysmus** Panz. ex Schult.

Mant. 2 (1824) 48.

1. Leaves cylindrical, somewhat ribbed, about 1 (2) mm broad. Glumes chestnut-brown. Nuts about 4 mm long; perianth bristles usually absent; if developed, shorter than nut.....2. **B. rufus** (Huds.) Link.
+ Leaves flat, up to 5 mm broad. Glumes rusty-brown. Nut about 2 mm long; perianth bristles present, 2-3 times longer than nut ... 2.
2. Perianth bristles erect or slightly contorted, thickened, 0.5-0.7 mm broad, twice longer than nut .... 1. **B. compressus** (L.) Panz. ex Link.
+ Perianth bristles highly contorted, especially in lower half, very thin, about 0.25 mm broad, almost thrice longer than nut ....3. **B. sinocompressus** Tang et Wang.

1. **B. compressus** (L.) Panz. ex Link, Hort. Berol. 1 (1827) 278; Rozhev. in Fl. SSSR, 3 (1935) 58, pro max. p.; Fl. Kirgiz. 2 (1950) 252; Fl. Kazakhst. 2 (1958) 22; Tang and Wang in Fl. R. P. Sin. 11 (1961) 40; Fl. Tadzh. 2 (1963) 51; Ikonnikov, Opred. rast. Pamira (Key to Plants of Pamir) (1963) 75. —*Schoenus compressus* L. Sp. pl. (1753) 43. —*Scirpus compressus* Pers. Syn. 1 (1805) 66; Ostenfeld in Hedin, S. Tibet (1922) 90; Persson in Bot. Notis (1938) 276. —**Ic.**: Fl. Tadzh. 2, Plate XI, figs. 1-2; Fl. R. P. Sin. 11, Table XVI, 5.

Described from Europe. Type in London (Linn.). Plate VIII, fig. 13.

In wet and marshy, sometimes saline meadows.

IB. Kashgar: ***North*** (southern margin of Mingbulak trough, 4-5 km south of Kein settlement, saline meadow, 1 IX; valley of Muzart river upper courses, Sazlik region, marshy meadow, 9 IX 1958—Yun.); West. ("Jerzil, in a meadow, about 2800 m, 1 VII 1930; Kentalek, about 2700 m, 14 VII 1931; Bostan-terek, swampy ground, about 2400 m, 14 VIII 1934"—Persson, l.c.).

IIA. Junggar: ***Altay region*** (in Qinhe region, in a gorge, No. 1528, 8 VIII 1956—Ching); ***Jung. Alt.*** (Toli district, on wet site, No. 2676, 6 VIII 1957—Kuan).

IIIC. Pamir ("Tash-korghan, about 3290 m, 28 VI 1935"—Persson, l.c.).

General distribution: Aral-Casp., Balkh. region, Jung.-Tarb., North and Centr. Tien Shan, East. Pamir; Europe, Near East, Caucasus, Mid. Asia.

2. **B. rufus** (Huds.) Link, Hort. Berol. 1 (1827) 278; Rozhev. in Fl. SSSR, 3 (1935) 58; Bohlin in Rep. Sci. Exped. N.-W Prov. China, S. Hedin, 11, 3 (1949) 28; Egorova in Bot. mater. Gerb. Bot. inst. AN SSSR, 19 (1959) 80; Fl. Tadzh. 2 (1963) 52; Ikonnikov, Opred. rast. Pamira (Key to Plants of Pamir) (1963) 75. —*Schoenus rufus* Huds. Fl. Angl. (1762) 15. —*Blysmus rufus* ssp. *exilis* Printz, Veg. Sib.-Mong. Front. (1921) 171. —*B. exilis* (Printz) Ivanova in Bot. zh. 24, 5-6 (1939) 502; Fl. Kirgiz. 2 (1950) 255; Fl. Kazakhst. 2 (1958) 23. —*Scirpus rufus* (Huds.) Schrad. Fl. Germ. 1 (1806) 133; Krylov, Fl. Zap. Sib. 3 (1929) 410; Pampanini, Fl. Carac. (1930) 81. —**Ic.**: Fl. Kirgiz. 2, Plate 46, fig. 2.

Described from Europe. Type in London (BM). Plate VIII, fig. 12.

In wet and marshy, often saline meadows, along banks of water reservoirs, near brooks, in swamps; up to 2700 m above sea level.

IA. Mongolia: ***Khobd., Mong. Alt., Cent. Khalkha*** (bank of Khaldangin-Nor lake on Mishik-Gun—Dol'che-Gegen road, VI 1926—Zam.); ***East. Mong.*** (near Kharkhonte railway station, solonchak, 8 VI 1902—Litw.); ***Depr. Lakes*** (south. bank of Khara-usu lake, swampy meadow solonchak, 20 VIII 1930—Bar.); ***Val. lakes*** (60 km west of Uburkhangai, meadow terrace of Shargin-Gol river, rather wet sedge grove, 15 VIII 1949—Kal.); ***Gobi-Alt.*** (Dundu-Saikhan hills, along bank of brook, 2 VII 1909—Czet.; southern trail of Artsa-Bogdo mountain belt, Dzhirgalant-khuduk collective, solonchak meadow, 20 VII 1948—Grub.); ***East. Gobi*** (Shabarakh-Usu, 1100 m, No. 87, 1925—Chaney); ***West. Gobi*** (Tsagan-Bogdo mountain range, Suchzhi-Bulak collective, saline meadow, 4 VIII 1943—Yun.); ***Alash. Gobi*** ("Edsen-Gol. dist. Bayan-bogdo, 1933—D. Hummel"—Bohlin, l.c.); ***Ordos*** (60 km west of Ushin town, bank of Ulibu-Nor salt lake, 2 VIII 1957—Petr.).

IC. Qaidam: ***mount.*** (Kurlyk area, in wet solonetz meadow, 2700 m, 10 V; Ikhe-Tsaidamin-Nor lake, in swamp, 3000 m, 14 VI 1895—Rob.).

IIA. Junggar: ***Jung. Gobi.***

IIIA. Qinghai: ***Nanshan*** (along Danhe river, 22 VII 1879—Przew.); ***Amdo*** (upper course of Huang He river, in swamps, 2400 m, 12 IV 1880—Przew.).

IIIC. Pamir: (Chumbus river, Kosh-Terek area, 7 VI 1909—Divn.).

General distribution: Jung.-Tarb., East. Pamir; Europe, Mid. Asia (West. Tien Shan), West. Sib. (south.), East. Sib. (south., excluding Sayan mountain range), North Mong. (Hang., Mong.-Daur.), Himalayas (west.).

3. **B. sinocompressus** Tang et Wang in Fl. R.P. Sin. 11 (1961) 224, 41. —*B. compressus* auct. non Panz. ex Link: Rozhev. in Fl. SSSR, 3 (1935) 58, p.p., quoad pl. Asiae centr.; Grubov, Consp. fl. MNR (1955) 80; Egorova in Bot. mater. Gerb. Bot. inst. AN SSSR, 19 (1959) 80. —*Scirpus caricis* auct. non Retz.: Hemsley, Fl. Tibet (1902) 201; Forbes and Hemsley, Index Fl. Sin. 3 (1905) 248; Strachey, Catal. (1906) 200. —*S. compressus* auct. non Pers.: Pampanini, Fl. Carac. (1930) 81. —*Nomochloa compressa* auct. non Bettle: Chen and Chou, Rast. pokrov r. Sulekhe (Plant Cover of Sulekhe River) (1957) 91. —**Ic.**: Fl. R. P. Sin. 11, Table XVI, 1-4.

Described from North-west. China (Gansu province). Type in Beijing. Plate VIII, fig. 14.

In solonchak and solonchak-like meadows, along banks of water reservoirs and near springs and brooks; up to 3200 m above sea level.

IA. Mongolia: ***Mong. Alt.*** (Tuguryuksk plain, valley of Tsinkir river, saline meadow, 15 VIII; lower Bodonchi river, ravine floor swamped by spring discharges, 1930—Bar.; Khasag-tu-Khairkhan hills, bank of Dundutseren-Gol river, 17 IX 1930—Pob.; valley of Bidzhi-Gol river, 5 km upstream from discharge of river onto foothill plain, rather saline meadow, 10 VIII 1947—Yun.); ***Cent. Khalkha*** (Ulan-Bator—Tsetserleg highway, sands along southern shore of Tsagan-Nur lake, saline grass plot, 25 VI 1948—Yun.); ***East. Mong., Val. lakes*** (valley of Ongiin-Gol river 40-45 km south of Khushu-Khid, hummocky meadow, 18 VII 1943—Yun.; Bain-Gobi somon, Tsagan-Gol river near somon camp site, grassland with sedge on river bank, 27 VII 1948—Grub.); ***Gobi-Alt.*** (35-40 km west-nor.-west of Dalan-Dzadagad, rather saline hummocky meadow, No. 10225, VII 1943—Yun.); ***Alash. Gobi*** (Bayan-Khoto, Tengeri sand, 2 VIII 1958—Petr.); ***Ordos*** (25 km south-east of Otok town, saline meadow near Khaolaitunao lake, 1 VIII; 60 km west of Ushin town, bank of saline lake Ulibu-Nor, 2 VIII 1957—Petr.); ***Khesi*** ("Sulekhe river"—Chen and Chou, l.c.).

IB. Kashgar: ***North*** (Engin, garden in Kara-tash floodplain, 2 VII 1929—Pop.; Aksu district, valley of Muzart river, saline meadow, No. 8287, 9 IX 1959—Lee et al.; valley of Gez-Darya river, 100 km from Kashgar, near hot spring in floodplain of river, 15 VI 1959—Yun.); South. (Keriya, along Nur river, swampy meadows, 23 VII 1885—Przew.; upper Cherchen river, 2700-3000 m, 9 VIII 1890—Rob.).

IC. Qaidam: ***plain*** (nor. slope of Burkhan-Budda mountain range, along Khatu-Gol river, along irrigation ditches, 21 VIII 1884—Przew.); ***montane*** (Baga-Tsaidamin-Nor lake, in meadow, 3000 m, 5 VI 1895—Rob.).

IIA. Junggar: ***Alt.*** (Toli-Myaoergou, near water, No. 2422, 4 VIII; along bank of Ven'tsyuan' river, near water, No. 1437, 14 VIII 1957—Kuan): ***Tien Shan*** (north of Kul'dzha, 3 V; Talla river, 18 VII; southern bank of Sairam lake, 1200 m, 20 VII; Dzhagastai, 900 m, VIII 1877—A Reg.; upper course of Taldy river, 19 VI; Mengute, 2700 m, 3 and 4 VII and 2 VIII 1879—A. Reg.): ***Jung. Gobi*** (south.: Gunlyu-Shakhe [Shikho], in gorge, near water, Nos. 3658 and 3676, 18 VIII 1957—Kuan; near Barkul' lake, No. 2231, 7 IX 1957—Kuan; east.: source of Ubchugiin-Gol river, near brook, 9 IX 1948—Grub.).

IIIA. Qinghai: ***Nanshan*** (Kuku-Nor lake, meadow along eastern shore, 3210 m, 5 VIII 1959—Pet.).

IIIB. Tibet: ***South.*** ("between Gunda-yaukti and Tazang, 3620-4800 m; Goring valley, 90°25′ E. long., 30°12′ N. lat., 4850 m—Littledale"—Hemsley, l.c.).

IIIC. Pamir (Chumbus river, Kosh-Terek area, 7 VI 1909—Divn.; valley of Tagarma river, marshy sires, 23 VIII 1913—Knorring; Issyk-su river estuary, 3100 m, 19 VII 1942—Serp.; valley of Tashkurgan river, 3000 m, No. 00296, 13 VI 1959—Lee et al.).

General distribution: Jung.-Tarb.; North Mong. (Hent., Hang., Mong.-Daur), China (North, North-west, South-west).

Note: The few known specimens from Kashgar, Pamir and partly Junggar (Tien Shan) differ from typical *B. sinocompressus* in somewhat more thickened perianth bristles approaching, in this feature, *B. compressus*. Compared with those of the latter, their bristles are, however, far more contorted and thinner.

10. Dichostylis Beauv.

in Lestib. Ess. Fam. Cypér. (1819) 39.

1. **D. hamulosa** (M.B.) Nees in Linnaea, 9 (1834) 289; Rozhev. in Fl. SSSR, 3 (1935) 63; Fl. Kazakhst. 2 (1958) 24. —*Cyperus hamulosus* M.B. Fl. taur.-cauc. 1 (1808) 35; Kük in Pflanzenreich, [IV, 20] 101 (1936) 502. —*Scirpus hamulosus* (M.B.) Stev. in Mém. Soc. natur. Moscou, 5 (1817) 356; Krylov, Fl. Zap. Sib. 3 (1929) 402. —**Ic.**: Fl. SSSR, 3, Plate V, fig. 5.

Described from Europe (lower Dnepr). Type in Leningrad.

IIA. Junggar: ***Jung. Gobi*** (nor.: Urungu river, 20 VIII 1876—Pot.).

Genera distribution: Aral-Casp., Balkh. region; Europe (south.), Asia Minor (plains), West. Sib. (south, excluding Altay).

11. Eleocharis R. Br.

Prodr fl. Hov. Holl. 1 (1810) 244.

1. Stylopodium (base of style) not separated from fruit apex but fused with it, poorly developed, triangular-acicular, brown, not spongy ....5. **E. meridionalis** Zinserl.
+ Stylopodium distinctly separated from fruit apex, very well developed, conical, spongy ....2.
2. Plants 2-10 (15) cm high, with very thin filiform stems. Spikelets small, 2-4 mm long and about 2 mm broad. Fruit oblong-obovate, about 1 mm long, with thin longitudinal veins and very thin transverse striations ....3.

+ Plants very large, stem not filiform. Spikelets 0.8-3 cm long, usually 3-5 mm broad. Fruit obovate, broad-ovate or subglobose, without veins and striations ....4.

3. Perianth bristles shorter, equal to or slightly longer than fruit, or absent ....1. **E. acicularis** (L.) Roem. et Schult.

+ Perianth bristles 1.5-2 times longer than fruits.................................. ....7. **E. yokoscensis** (Franch. et Savat.) Tang et Wang.

4(2). Spikelet with single glume at base....6. **E. uniglumis** (Link) Schult.

+ Spikelet with two glumes at base..5.

5. Spikelet not dense, fruits few, with brownish-purple glumes, without white membranous margin or with very narrow one. Stem thinly striate, not veined. Perianth bristles usually achromatic...... ....4. **E. intersita** Zinserl.

+ Spikelet very dense, fruits many, glumes with very broad white membranous margin, quite often almost wholly silvery-white, with only narrow, coloured striations along central light-coloured vein. Stem usually ribbed; only large specimens have thick stems, latter subglabrous. Perianth bristles brown.....................................6.

6. Spikelets (1) 1.5-3 cm long, narrowly conical, usually long and acuminate. Glumes elongate-lanceolate, acuminate or acute, silvery, with very narrow brownish striations only along light-coloured central band and very broad white membranous margin. Perianth bristles thin, mostly shorter than nut, sometimes absent, very rarely slightly larger than nut; teeth on bristles barely developed..2. **E. argyrolepis** Kier. ex Bge.

+ Spikelets 0.8-1 cm, less often 1.5 cm long, oblong- or elongate-ovate, acute or obtuse, not long and acuminate. Glumes oblong-ovate or elongate-ovate, obtuse or acute, with fairly broad purplish-brown bands with broad white membranous margin along light-coloured midrib. Perianth bristles invariably present, longer than nut, with well-developed teeth..................................... ....3. **E. equisetiformis** (Meinsh.) B. Fedtsch.

1. **E. acicularis** (L.) Roem. et Schult. Syst. 2 (1817) 154; Krylov, Fl. Zap. Sib. 3 (1929) 394; Zinserl. in Fl. SSSR, 3 (1935) 70; Grubov, Consp. Fl. MNR (1955) 80; Fl. Kazakhst. 2 (1958) 26. —*E. yokoscensis* (Franch. et Savat.) Tang et Wang in Fl. R. P. Sin. 11 (1961) 54, p.p., quoad pl. songor. chin. —*Scirpus acicularis* L. Sp. pl. (1753) 48. —**Ic.**: Fl. SSSR, 3, Plate VI, fig. 4.

Described from Europe. Type in London (Linn.).

IA. Mongolia: ***Khobd., Mong. Alt., East. Mong.***

IIA. Junggar: ***Alt. region, Jung. Gobi*** (Ch. Irtysh [near Dyurbel'dzhina], 26 VIII 1876—Pot.).

General distribution: Aral-Casp., Balkh. region; Arct., Europe, Caucasus, West. Sib. (predominantly south.), East. Sib., North Mong. (Hent., Hang., Mong.-Daur.), North Amer.

2. **E. argyrolepis** Kier. in Bge. Beitz. Kenntn. Fl. Tussl. (1852) 342; idem in Mém. Sav. Étr. Ac. Sci. Pétersb. 7 (1854) 518; Zinserl. in Fl. SSSR, 3 (1935) 79; Persson in Bot. Notis. (1938) 276; Fl. Kirgiz. 2 (1950) 256; Fl. Kazakhst. 2 (1958) 30; Tang and Wang in Fl. R. P. Sin. 11 (1961) 64; Fl. Tadzh. 2 (1963) 56. —**Ic.**: Fl. Kirgiz. 2, Plate 47, fig. 2.

Described from Mid. Asia (Syr-Darya river basin). Type in Leningrad. Plate VIII, fig. 1.

In water, on marshy sites, sink holes, saline meadows; from foothills to 2400 m above sea level.

IB. Kashgar: ***North*** (Yarkand-Darya, among reeds, on clay, 15 VI 1889—Rob.; Karashar district, south-east of "Khugan" state farm, in water, 1100 m, No. 6087, 25 VII; Yuili, in Lobulok lake, in water, Nos. 8570 and 8604, 8 and 12 VIII 1958—Lee and Chu; "Maralbashi, in a swamp, about 1070 m, 14 VI 1932"—Persson, l.c.); West. ("Bostan-terek, in a ditch, about 2400 m, 14 VIII 1934"—Persson, l.c.).

IIA. Junggar: ***Jung. Gobi*** (north-west: valley of Darbaty river near Karamai-Altay highway intersection, reed thickets along margin of solonchak, 20 VI; Orkhu settlement on Karamai-Altay highway, near Dyan river intersection, saline meadow, 21 VI 1957—Yun.; south.: along bank of Manas river, on marshy sites, No. 794, 11 VI; Savan district, Seyadi, No. 1504, 16 VI; same site, Shamyntsza, No. 955, 17 VI 1957—Kuan; lower course of Manas river, 55 km nor.-nor.-west of "Paodai" state farm, saline meadow, 17 VI 1957—Yun; Usu district, San'tszyao-chzhuan, on west site, No. 1057, 25 VI; Savan district, Mogukhu water reservoir, No. 1563, 25 VI 1957—Kuan).

General distribution: Aral-Casp., Balkh. region. ?Jung.-Tarb., North and Centr. Tien Shan; ?Caucasus (Cis-Caucasus), Mid. Asia (plains, West. Pamiroalay and West. Tien Shan).

3. **E. equisetiformis** (Meinsh.) B. Fedtsch. in Rast Turkest. (Plants of Turkestan) (1915) 165; Zinserl. in Fl. SSSR, 3 (1935) 80; Fl. Kirgiz. 2 (1950) 256; Fl. Kazakhst. 2 (1955) 30; Fl. Tadzh. 2 (1963) 58. —*E. valleculosa* Ohwi in Acta Phytotax. and Geobot. 2 (1933) 29; Tang and Wang in Fl. R. P. Sin. 11 (1961) 65. —*Scirpus equisetiformis* Meinsh. in Acta Hort. Petrop. 18 (1901) 261. —**Ic.**: Fl. SSSR, 3, Plate VI, fig. 20; Fl. R. P. Sin. 11, Plate XXIII, figs. 12-14 (sub *E. valleculosa*).

Described from China (Kul'dzha). Type in Leningrad. Plate VIII, fig. 2.

On lake shores, in water, grassy swamps fed by springs, and stagnant water.

IA. Mongolia: ***Mong. Alt., Alash. Gobi, Ordos*** (valley of Huang He river, 1 VIII 1871—Przew.; Termin-Bashin area, 9 VIII 1884—Pot.; 10 km south-west of Ushin town, edge of a pond, 4 VIII; 20 km west of Dzhasak town, in a valley, 16 VIII; 50 km south of Dzhasak town, meadow with puddle among sand dunes, 17 VIII 1957—Petr.); ***Khesi*** (Edzin river above and below Gaotai, in stagnant water, 20 VI 1886—Pot.).

IB. Kashgar: ***East.*** (Khami, in water, 29 V 1877—Pot.).

IIA. Junggar: ***Tien Shan*** (north of Kul'dzha, 3 V 1877—A. Reg.; 20 km south-west of Urumchi, No. 0428, 19 VII 1956—Ching); ***Jung. Gobi*** (south.: 3-4 km east of St. Kuitun settlement on Shikho-Manas road, Konkada area, grassy swamp fed by a spring, 30 VI, 6 and 7 VII 1957—Yun.; Kuitun region, in meadow, No. 255, 29 VI; 2 km nor. of Kuitun, No. 389, 6 VII; 3 km south of Kuitun, No. 433, 7 VII 1957—Kuan); ***Dzhark.*** (Kul'dzha, 1876—Golike; same site, 15, V, VII, VIII 1877—A. Reg.); ***Balkh.-Alak.*** (south of Dachen, in swamp, No. 2877, 11 VIII 1957—Kuan).

General distribution: Aral-Casp., Balkh. region, Jung.-Tarb., North Tien Shan; Mid. Asia (West. Pamiroalay, West. Tien Shan), China (widespread).

4. **E. intersita** Zinserl. in Fl. SSSR, 3 (1935) 581, 76; Kitag. Lin. Fl. Mansh. (1939) 119; Fl. Kazakhst. 2 (1958) 27; Tang and Wang in Fl. R. P. Sin. 11 (1961) 67. —?*E. mamillata* Lindb. f. var. *cyclocarpa* Kitag. l.c. 119; Tang and Wang, l.c. 64. —*E. eupalustris* auct. non Lindb. f.: Grubov , Consp. fl. MNR (1955) 80, p.p. —*E. palustris* auct. non R. Br.: Forbes and Hemsley, Index Fl. Sin. 3 (1905) 227, p.p.; Danguy in Bull. Mus. nat. hist. natur. 17, 6 (1911) 451; Chen and Chou, Rast. pokrov r. Sulekhe (Plant Cover of Sulekhe River) (1957) 81. —*Scirpus palustris* auct. non L.: Franch. Pl. David. 1 (1884) 317. —**Ic.**: Fl. SSSR, 3, Plate VI, figs. 11 and 26.

Described from Europe (Belorussia). Type in Leningrad.

Along banks of water reservoirs and in shallow water, grassy marshes, near springs and brooks, in marshy floodplains, often solonetz and saline meadows; from plains and foothills to 2460 m above sea level.

IA. Mongolia: ***Khobd., Mong. Alt.*** (floodplain of Bulugun river at the point of inflow into it of Uliste-Gol from left, 20 VII 1947—Yun.); ***Cis-Hinggan*** (Yaksha railway station, No. 2188, 1954—Wang; Khamar-Daban region, floodplain of Khalkhin-Gol river, 19 VI 1954—Dashnyam); ***Cent. Khalkha*** (valley of Kerulen river, 11 VII; Khurkhu river, 23 VII 1894—Kashk.; Orkhon-Gol river, near Erdeni-dzu monastery on left bank, 21 VI; south. shore of lower Tsagan-Nur lake on Ulan-Bator—Tsetserleg highway, small freshwater ponds, 25 VI 1948—Grub.); ***East. Mong.*** (Khailar railway station, 6 VII 1901—Lipsky; Khailar town, Nunlin'tun' village, near water, No. 627, 10 VI 1951—Wang; Khailar town, solonetz water reservoir, Nos. 2980 and 3004, 29 VII 1954—Wang); ***Depr. Lakes, Val. Lakes*** (west. shore of Orok-Nur lake, saline meadow, 2 VII 1941—Tsatsenkin); ***Gobi-Alt.*** (Dzhirgalante-Bulak spring south of Noin-Bogdo mountain range, 23 VII 1929—Glag.; 35-40 km west-nor.-west of Dalan-Dzadagad, No. 10222, VI 1943—Yun.); ***East. Gobi*** (Alashan-Urgu road, valley of Urten-Gol river, No. 341, 22 VII 1909—Czet.; Under-shili somon, Toli-bulak area, saline meadow near brook, 27 VII 1946—Yun.); ***West. Gobi*** (30-35 km east of Tsagan-Bogdo, saline meadow near spring, 31 VII; Tseel' somon, Ikhe-Tszaram area south of Mongolian Altay, marshy saline meadows, 19 VIII 1943—Yun.); ***Alash. Gobi*** (Mukhur-shanda spring on border road to south of Tostu-nur mountain range, 13 VIII 1948—Grub.; Bayan-Khoto, Tengeri sands, Turgu-Nor, 14 VIII 1958—Petr.); ***Khesi*** (between Gaotai and Fuiitin, 12 and 18 VI 1886—Pot.; "Lac de Je-Jue-Ts'iuan, region des dunes, environs de Cha-tscheou [Sachzhou], No. 689; 4 VI 1908—Vaillant"—Danguy, l.c.; "r. Sulekhe"—Chen and Chou, l.c.).

IB. Kashgar: ***North*** (Yuili district, Lobulok lake, in water, No. 8567, 8 VIII 1958—Lee and Chu; lower course of Tarim river, 3-5 km nor.-west of Char settlement, marshy saline meadow on floodplain, 19 VIII 1958—Yun.); ***West.*** (Ulugchat-Baikurt, 2200 m, No. 09693, 19 VI; north of bridge on Sufu [Kashgar]-Barkhetai highway, No. 00017, 22 VI 1959—Lee et al.); ***South.*** (right bank of Kerii river, in floodplain, No. 00119, 16 V 1959—Lee et al.; 10 km north of Kerii town, Bostan area, marshy meadow, 14 VI; 8 km west of Kerii, on Chira road, marshy meadow, 16 VI 1959—Yun.; 6 km nor.-west of Cherchen district, No. 9507, 12 VI; Niya district, 1393 m, No. 9565, 23 VI 1959—Lee et al.); ***East.*** (in Turfan region, Yaer lake, near water, No. 5562, 9 VI; nor.-east of Toksun, marsh, No. 7318, 19 VI 1958—Lee and Chu); ***Takla-Makan*** (valley of Niya river 20 km nor.-east of Niya settlement, sedge meadow in rather saline floodplain, 7 V 1959—Yun.).

IIA. Junggar: ***Altay region*** (in Qinhe region, 2000 m, No. 1534, 8 VIII; in Koktogoi region, 1200 m, No. 198, 18 VIII 1956—Ching; Koktogoi, near water, 950 m, Nos. 10403 and 10407, 7 VI 1959—Lee et al.); ***Tien Shan*** (B. Yuldus, on marsh, 2460 m, No. 6482, 10 VIII 1958—Lee and Chu); ***Jung. Gobi*** (nor.: Ch. Irtysh, 26 VIII 1876—Pot.; south.: in Ulausu region, saline marshy meadows, No. 318, 3 VII; 3 km east of Kuitun, on marsh, No. 434, 7 VII 1957—Kuan; same site, grassy marsh fed by spring, 7 VII 1957—Yun.).

General distribution: Arct. (Asiat.), Europe, Caucasus, West. Sib., East. Sib., Far East, North Mong. (Hent., Hang.), China (Dunbei, North, South-west.), Korean peninsula, North Amer.

Note: The distribution of this species outside Central Asia has not been adequately studied because of its confusion with *E. palustris* (L.) R. Br. *E. intersita* differs from the latter in stylopodium most often in the form of an equilateral triangle in longitudinal section as well as in the generally white perianth bristles.

Plants from Junggar Gobi (Kuituna district) differ from typical *E. intersita* in the shape of the stylopodium, which is significantly broader than long.

5. **E. meridionalis** Zinserl. in Fl. SSSR, 3 (1935) 580, 69; Fl. Kirgiz. 2 (1950) 255; Fl. Kazakhst. 2 (1958) 26; Fl. Tadzh. 2 (1963) 55; Ikonnikov, Opred. rast. Pamira (Key to Plants of Pamir) (1963) 77. —*E. pauciflora* auct. non Link: Grubov, Consp. fl. MNR (1955) 80; Tang and Wang in Fl. R. P. Sin. 11 (1961) 52. —*Scirpus pauciflorus* auct. non Lightf.: Persson in Bot. Notis. (1938) 276. —**Ic.**: Fl. SSSR, 3, Plate VI, fig. 1; Fl. R. P. Sin. 11, + XX, 1-6 (sub *Eleocharis pauciflora*).

Described from Pamiroalay (Alay mountain range). Type in Leningrad.

Saline grassland near springs, in water, on swamps fed by spring, on marshy sites along banks of brooks; from foothills to 1300-2800 m above sea level.

IA. Mongolia: ***Mong. Alt.*** (Khasagtu-Khairkhan hills, Dundu-Tseren-Gol river near Undur-Khairkhan, 16 IX 1930—Pob.); ***Cent. Khalkha*** (15 km nor.-west of Saikhan-obo somon, along valley of Ongiin-Gol river, saline meadow, 9 VII 1941—Tsatsenkin); ***Bas. lakes, West. Gobi*** (south. foothills of Tsagan-Bogdo mountain range, Tsagan-Bulak area, saline short-grass meadow along spring, 1 VIII 1943—Yun.); ***Ordos*** (20 km west of Dzhasak town, in a valley, 16 VIII 1957—Petr.).

IB. Kashgar: ***North*** (south. slope of Keinsk trough in upper course of Kyzyl river, north of Kucha town, saline meadow near spring, 2 IX 1958—Yun.); ***West.*** (valley of Gez-Darya river, 100 km from Kashgar on road to Bulunkul' and Tashkurgan, in water and along bank of hot spring, 15 VI 1959—Yun.; "Kentalek, in a meadow, about 2700 m, 12 VII 1931; Kashgar, in a swamp about 25 km south of town, about 1300 m, 12 V 1935"—Persson, l.c.).

IIA. Junggar: ***Jung. Alt.*** (between Sumbe and Kasan, 2100-2400 m, 22 V 1878—A. Reg.); ***Tien Shan*** (Kul'dzha, 1876; north of Kul'dzha, 3 V; Toguztarsu near Kul'dzha, 5 V; Sairam, 1200-2700 m, VII 1877; Mengute, 13 VII 1879—A. Reg.; on B. Yuldus swamp, near water, 2460 m, No. 6469, 10 VIII 1958—Lee and Chu; B. Yullus basin 30-35 km south-west of Bain-Bulak settlement, along swamps fed by springs, 10 VIII 1958—Yun.).

IIIB. Tibet: ***Chang Tang*** (Muzlyk-atasy mountain range, nor. slope, near springs [1890]—Rob.).

IIIC. Pamir (Chumbus river, Kosh-Terek area, 7 VI 1909—Divn.; Tagarma valley, marshy sites, 23 VII 1913—Knorring; Tashkurgan district, in a valley, No. 00295, 13 VI: Bulunkul', 44 km from Gez-Darya river, 2800 m, No. 00364, 15 VI 1959—Lee).

General distribution: Jung.-Tarb., Nor. and Cent. Tien Shan, East. Pam.; Caucasus (South. and Transcaucasus), Mid. Asia (West. Tien Shan, Pamiroalay), North Mong. (Mong.-Daur.).

Note: This species differs quite distinctly from the related boreal and predominantly Eurasian species *E. pauciflora* (Lightf.) Link in very small fruit—1.8-2.1 (2.5) mm long and 0.9-1 mm broad (2.5-2.7 mm long and 1.2-1.4 mm broad in *E. pauciflora*)—as well as very short stylopodium, which is almost wholly blackish-brown in ripe fruit.

6. **E. uniglumis** (Link) Schult. Mant. 2 (1824) 88; Franch. Pl. David. 1 (1884) 318; Persson in Bot. Notis. (1938) 276; Grubov, Consp. fl. MNR (1955)

80; Tang and Wang in Fl. R. P. Sin. 11 (1961) 68; Fl. Tadzh. 2 (1963) 58; Ikonnikov, Opred. rast. Pamira (Key to Plants of Pamir) (1963) 77. —*E. palustris* var. *uniglumis* (Link) Schmalh. in Fl. Sr. i Yuzhn. Ross. 2 (1897) 542; Krylov, Fl. Zap. Sib. 3 (1929) 393. —*E. euuniglumis* Zinserl. in Fl. SSSR, 3 (1935) 82; Fl. Kirgiz. 2 (1950) 256; Fl. Kazakhst. 2 (1955) 30. —*E. palustris* auct. non R. Br.: Forbes and Hemsley, Index Fl. Sin. 3 (1905) 227, p.p. —*Scirpus uniglumis* Link in Spreng. Jahrb. 3 (1820) 77. —**Ic.**: Fl. Tadzh. 2, Plate VI, figs. 8 and 9.

Described from Europe. Type in Berlin.

Along river banks in water, near springs and wells, in ponds and lakes, swamps and marshy saline meadows; from foothills to 2500-3000 m above sea level.

IA. Mongolia: ***Cent. Khalkha*** (Kerulen river near San-beise stop, in meadows, 1899—Pal.); ***East. Mong., Depr. Lakes*** (Shargain-Gobi, Tongulak-Bulak spring, among puffed solonchak, 5 IX 1948—Grub.); ***Gobi Alt.*** (Bain-Tukhum area, 4 VIII 1931—Ik.-Gal.); ***East. Gobi*** (Shabarakh-usu [Bain-tszak], in marshes, 1100 m, No. 81, 1925—Chaney); ***West. Gobi*** (south. foot of Tsagan-Bogdo mountain range, saline meadow near spring, 1 VIII 1943—Yun.); ***Alash. Gobi*** (Fyn'-yuan' in oasis [Bayan-Khoto], on marshy sites, 14 IV 1908—Czet.); ***Khesi*** (between Shakhe and Fuiitin villages, 5 VI; between Gaotai and Fuiitin, 6 and 12 VI; Kheikho river below Gaotai, 20 VI 1886—Pot.; nor. foothills of Nanshan, Dauchan', 1500 m, along springs, 12 V 1894—Rob.; 40 km north of Yunchan town, Ninyanlu settlement, wet meadow in valley, 11 VII 1958—Petr.).

IB. Kashgar: ***North*** (Yarkand-Darya, among rushes, 15 VI 1889—Maralbashi, Chuderlik village, loess, 6 V; Uchturfan, 14 V 1908—Divn.; Pichan intermontane basin, Chon-Karadzhal settlement, region of solonchak troughs, near spring, 22 VI 1959—Yun.; 1 km north of Pichan district, near water, 330 m, No. 5428, 23 V; Pichan district, 100 km west of Khando, in a pond, 460 m, No. 6667, 13 VI; 70 km from Faizabad on road to Maralbashi, Bachu, No. 7553, 29 IX 1958—Lee and Chu); ***West.*** (10 km south of Upal oasis on road to Tashkurgan, sasa zone, near spring, 11 VI 1959—Yun.; "Kashgar, about 1330 m, 12 V 1935"—Persson, l.c.); ***South.*** (Niya oasis, 1260 m, 3 VI 1885—Przew.); ***East.*** (Ledun in Khami region, in water, No. 462, 21 V 1957—Kuan; in Turfan region, Putougou, near water, 170 and 100 m, Nos. 5495 and 5506, 1 VI 1958—Lee and Chu); ***Takla-Makan*** (20 km north-east of Niya, on floodplain, No. 00048, 7 V 1959—Lee et al.).

IC. Qaidam: ***plains*** (Syrtyn valley, Ikhen-shirik area, in marsh, 3000 m, 20 VII 1895—Rob.); ***montane*** (Kurlyk area, along wet saline meadow, 2700 m, 15 V 1895—Rob.).

IIA. Junggar: ***Tien Shan*** (in Urumchi region, No. 590, 2 VI 1957—Kuan; B. Yulduus basin, 30-35 km south-west of Bain-Bulak settlement, marshy meadow, 10 VIII 1958—Yun.); ***Jung. Gobi*** (south.: Savan district, Mogukhu lake, along shore, No. 1554, 20 VI 1957—Kuan).

IIIA. Qinghai: ***Amdo*** (upper Huang He, 2400 m, 2 VI 1880—Przew.).

IIIC. Pamir: (Chumbus river, Kosh-Terek area, 7 V 1909—Divn.).

General distribution: Aral-Casp., Balkh. region, North and Centr. Tien Shan, East. Pam.; Arct., Europe, Near East, Caucasus, Mid. Asia, West. Sib. and East. Sib. (predominantly south.), Far East (Kamchatka), North. Mong. (Hang., Mong.-Daur.), China (North.-west.), ?Himalayas.

7. **E. yokoscensis** (Franch. et Savat.) Tang et Wang in Fl. R. P. Sin. 11 (1961) 54, p.p. excl. pl. songor. chin. —*Scirpus yokoscensis* Franch. et Savat. Enum. Pl. Japon. 2 (1879) 543, 109. —*E. svensonii* Zinserl. in Fl. SSSR, 3 (1935) 580, 71; Grubov, Consp. fl. MNR (1955) 80. —*E. acicularis* auct. non (L.) Roem. et Schult.: Forbes and Hemsley, Index Fl. Sin 3 (1905) 225. —**Ic.**:

Fl. SSSR, 3, Plate VI, fig. 5 (sub *E. svensonii*); Fl. R.P. Sin. 11, Table XX, 10-13.

Described from Japan (Honshu island). Type in ?Paris. Isotype in Leningrad.

In shallow water.

IA. Mongolia: ***East. Mong.*** (near Khailar town, in shallow water, No. 755, 16 VI 1951—S. H. Li; "East. Mong. - north"—Grubov, l.c.).

General distribution: Far East (south.), North Mong. (Hent., Hang., Mong.-Daur.), China, Korea, Japan.

12. **Kobresia** Willd.

Sp. pl. 4 (1805) 205.

1. Plant dioecious. Pistillate spikelets linear, green. Perigynium closed like utricle in *Carex* .... 17. **K. prattii** Clarke.
+ Plant monoecious. Perigynium not fully closed or, less often, 2/3 of it closed .... 2.
2. Perigynium not closed, with margins connate only in lowermost portion. Glume and perigynium less than 10 mm long................ 3.
+ Perigynium closed with margins connate up to 2/3. Glume and perigynium about 10 mm long .... 8. **K. robusta** Maxim.
3. Shoots surrounded at base by several sheaths of old leaves, usually with fallen (as though sheared) blades. Leaves erect, usually narrow and convoluted like a bristle. Stems almost cylindrical. Nut generally 1.5-2 times longer than broad and mostly with short beak, often with browning remnant of stylopodium..................... 4.
+ Blades usually preserved in sheath surrounding shoot base. Leaves not erect, often curved, flat or longitudinally folded, sometimes very narrow but then flattened laterally. Stems generally trigonous. Nut mostly (2) 2.5-4 times longer than broad. Nut usually with non-browning elongated remnant of stylopodium-beak (less often without stylopodium—*K. humilis*) .... 10.
4. Roots thickened with dense hairs. Stems thickened, (1) 1.5-2 (3) mm in diameter. Sheath surrounding base of shoots, (3) 5-10 mm broad. Nut lustrous or matte .... 5.
+ Roots slender, glabrous. Stems slender, sometimes filiform, 0.5-1.2 mm in diameter. Sheath surrounding shoot base narrow, 1.5-3.5 mm broad. Nut lustrous .... 6.
5. Inflorescence—simple spike—externally very similar to corresponding ones in members of genus *Eleocharis*. Nut lustrous .... 7. **K. tibetica** Maxim.
+ Inflorescence compound, spicate. Nut matte.................................... .... 6. **K. smirnovii** Ivanova.
6. Inflorescence compound, spicate...... 4. **K. filifolia** (Turcz.) Clarke.
+ Inflorescence—simple spike .... 7.

7. Sheath surrounding shoot base usually light or pale brown. Leaves soft, filiform. Spikelets biflorous, with one pistillate and one staminate flower. Glumes 2.5-3 mm long. Nut (without beak) 2 mm long .... 1. **K. bellardii** (All.) Degl.
+ Sheath surrounding shoot base usually dark chestnut-brown or blackish-brown. Spikelets with one pistillate and 2 or 3 (4) staminate flowers. Glume 4.5-6 mm long. Nut (without beak) 2.5-3.5 mm long .... 8.
8. Glume usually chestnut-brown, without white membranous margin. Perigynium (2) 2.5-3.2 (3.5) mm long, separated from inflorescence axis with difficulty (nut shed without perigynium). Nut (without beak) 2.5-2.7 (3) mm long.... 3. **K. capilliformis** Ivanova.
+ Glume brownish or light brown, usually with white membranous margin. Perigynium (3.8) 4-5 (5.5) mm long, readily separated from inflorescence axis (shed along with nut). Nut (without beak) 2.5-3.5 mm long .... 9.
9. Nut (without beak) 2.5–2.7 (3) mm long ... 5. **K. ovczinnikovii** Egor.
+ Nut (without beak) 3.3–3.5 mm long... .... 2. **K. capillifolia** (Decne.) Clarke.
10 (3). Inflorescence—simple spike, sometimes with 1 or 2 short, poorly distinct branches in lower portion; 0.5–1.5 (2) cm long and 0.4–1 cm broad. Plant 1–15 cm high... 11.
+ Inflorescence compound, branched, spicate, spicate-paniculate or head ovate or oblong-ovate .... 14.
11. Plant 1–2 cm high. Leaves about 1 mm broad. Perigynium unisexual with only a single pistillate flower............. 13. **K. pygmaea** Clarke.
+ Plant very tall. Leaves 1–2.5 mm broad. Perigynium bisexual ...... 12.
12. Stigmas three. Plant found in Junggar, Kashgar and Pamir ...... 13.
+ Stigmas 2 or 3. Plan found in Tibet............ 12. **K. pusilla** Ivanova.
13. Spike rusty-brown. Nut broadly oribculate above, with stylopodium browning down to base and shedding... .... 9. **K. humilis** (C.A. Mey. ex Trautv.) Serg.
+ Spike usually chestnut-brown. Nut conical above due to unbrowned remnant of stylopodium.... 11. **K. persica** Kük. et Bornm.
14 (10). Inflorescence capitate or slightly lobed, 1–1.5 (2) cm long. Lower half of stem 0.7–1.2 mm in diameter. Leaves 1–2 mm broad, longitudinally folded .... 15. **K. simpliciuscula** (Wahl.) Mack.
+ Inflorescence (2) 3–5 cm long. Lower half of stem 2.5–3 mm in diameter. Leaves (2) 3–7 mm broad, flat or longitudinally folded................ 15.
15. Inflorescence very dense, compact, cylindrical. Stem under inflorescence 2 mm in diameter. Leaves equal to or slightly shorter than stem, suberect. Perigynium with broad orbicular apex. Roots with dense hairs .... 10. **K. knsuensis** Kük.

+ Inflorescence mostly somewhat lax and lobed, spicate-paniculate. Stem under inflorescence 1 mm in diameter. Leaves 1.5–2 times shorter than stem, generally drooping. Perigynium constricted toward top. Roots glabrous ....16.

16. Nut with parallel sides and long beak, 0.7–0.9 mm in length. Margins of perigynium in mature nut closed almost up to very apex. ....16. **K. stenocarpa** (Kar. et Kir.) Steud.

+ Nut constricted toward both ends, with very short beak, 0.4 mm long. Margins of perigynium in mature nut closed in lower portion. ....14. **K. royleana** (Nees) Boeck.

Subgenus **Elyna** (Schrad.) Ivanova

Section Elyna (Schrad.) Ivannova

1. **K. bellardii** (All.) Degl. in Loisel. Fl. Gall. 2 (1807) 626; Kük. in Pflanzenreich [IV, 20], 38 (1909) 37; Krylov, Fl. Zap. Sib. 3 (1929) 413; Sergievsk. in Fl. SSSR, 3 (1935) 109, p.p.; Ivanova in Bot. zh. 24: 5–6 (1939) 486; Grubov, Consp. Fl. MNR (1955) 81; Fl. Kazakhst. 2 (1958) 37. —*K. myosuroides* (Vill.) Fiori et Paol. Ic. Fl. Ital. (1895) 52; Kitag. Lin. Fl. Mansh. (1939) 121. —*Carex bellardii* All. Fl. Pedem. 2 (1785) 264. —**Ic.:** Fl. SSSR, 3, Plate IX, fig. 7.

Described from Europe (Alps). Type in ?. Plate VII, fig. 6; map 1.

In meadows, on mountain steppes, turf-covered rocky slopes, rubble placers; in alpine and subalpine belts, less frequently in larch forests.

IA. Mongolia: ***Khobd.*** (Bairimen-daban, hill top, 20 VI 1879—Pot.); ***Mong. Alt.*** (Taishiri-ula mountain range, north slope, larch forest, 12 VII 1945—Yun.; same site, 21 IX 1945, Leont'ev; same site, 15 km south-east of Yusun-Bulak, meadow on fringe of larch forest, 1 IX 1948—Grub.; Tolbo-Kungei-Alatau mountain range 3200 m, high-altitude belt, Kobresia meadow, 5 VIII 1945—Yun.; Adzhi-Bogdo mountain range, Burgasin-dada pass, rubble placers of alpine belt, 6 VIII; valley of Bidzhi-Gol river, saline meadow with birches, 10 VIII 1947, Yun.); ***Depr. Lakes, Cis-Hing.*** (Arshan town, dry site on hill meadow, No. 315, 14 VI 1950—Chan); ***Gobi-Alt.*** (Dzun-Saikhan hill, on dry meadow at top of ravine, 23 VIII 1931—Ik.-Gal.; Baga Bodgo, in gravel at stream edge at 1800 m, No. 247, 1925—Chaney; Ikhe-Bogdo mountain range, south. slope, upper part of Narin-Khurimt ravine, Kobresia steppe, 5 IX 1943—Yun.; same site, fescue-sedge steppe, 28 VI; same site, above Ketsu ravine, Kobresia thicket, 29 VI 1945—Yun.; same site, streams feeding Narin-Khurimt gorge, 3500 m, site among rocks, 29 VII 1948—Grub.; nor. slope of Ikhe-Bogdo mountain renge, upper Shishkhid ravine, hill steppe, June 30; Tsagan-Gol river, along bank, 3 VII 1945—Yun.).

IIA. Junggar: ***Alt. Region*** (Qinhe district, Chzhunkhaitsza, 2460 m. No. 1386, 6 VIII 1956—Ching; Altay hills in Timulbakhan region, in high-altitude belt, 2600 m, No. 10702, 19 VII 1959—Lee et al.), ***Jung, Gobi, Tarbag.*** (south. slope of Saur mountain range, in Khobuk river valley, Karagaitu gorge, high-altitude belt, kobresia meadows, 23 VI 1957—Yun et al.); ***Tien Shan*** (Aryslyn, 16 VII 1879—A. Reg.; Turkul' lake, on marshy site, 17 VI 1877—Pot.; left bank of Manas river, upper course of Danu-Gol river as it rises into Sedaban pass, high-altitude belt, grassy-Kobresia meadow on semi-turfed talus, 21 VII 1957—Yun.; along Danu-Gol river 2800 m, No. 2083, 21 VII Bargat, off Yakou in north, No. 1732, 31 VIII 1957—Kuan).

IIIA. Qinghai: ***Nanshan*** (108 km-west of Xining, hilly scrub steppe, 3400 m, 5 VIII; Kuku-Nor lake, meadow along west. bank of lake, 3210 m, 5 VIII 1959—Petr.).

General distribution: Jung-Tarb.; Arct., Europe (mountains of West. Europe, Urals), West. Sib. (Altay), East. Sib., Far East (Kuril islands), North. Mong. (Fore Hubs., Hent., Hang.), China (Dunbei), Korea, Japan, North Amer. (Rocky Mountains).

2. **K. capillifolia** (Decne.) Clarke in J. Linn. Soc. (London) Bot. 20 (1883) 378; Kük. in Pflanzenreich [IV, 20], 38 (1909) 36, p.p.; Ivanova in Bot. zh. 24: 5–6 (1939) 486, p.p. (excl. *K. macrolepis* Meinsh.); Pampanini, Fl. Carac. (1930) 82. —*Elyna capillifolia* Decne. in Jacquem. Voy. Ind. Descr. Coll. Bot. (1844) 173.

Described from West. Himalayas. Type in Paris.

In meadows, on rocky sites and along river banks in high-altitude belt.

IIIB. Tibet: ***Chang Tang*** (Keriya mountain range [Russky] valley of Kyuk-Egil' river, in rocky sites, 3750 m, 11 VII 1885—Przew.); ***Weitzan*** (left bank of Dychu river [Yangtze] in midcourse, on cliffs, 3900 m, 23 VI 1884—Przew.; north slope of Burkhan-Budda mountain range, Nomokhun gorge, 3600–4200 m, 18 V 1900 Lad.).

General distribution: Near East (Afghanistan, Pakistan—Hindukush); China (South-west), Himalayas (west., Kashmir).

Note. The above herbarium specimen from Chang Tang was collected in an early stage of development, i.e., non-flowering state, and hence could not have been accurately described. Following N.A. Ivanova, we placed it among *K. capillifolia* although it can be identified with *K. capilliformis* from the colour of glumes (chestnut-brown, without white membranous margin) and size of perigynium (3.5 mm long).

Causasian plants described as *K. macrolepis* Meinsh. and regarded as a synonym for *K. capillifolia* by N.A. Ivanova (l.c.) should evidently be treated as a distinct race. They differ from *K. capillifolia* in colour of the glume (dark chestnut-brown, without white membranous margin among *K. macrolepis* and rather light brown or nearly chestnut-brown but with white membranous margin among *K. capillifolia*).

3. **K. capilliformis** Ivanova in Bot. zh. 24: 5–6 (1939) 484; Fl. Kirgiz. 2 (1950) 262; Fl. Kazakhst. 2 (1958) 36. —*K. capillifolia* ssp. *capilliformis* (Ivanova) Ovcz. in Fl. Tadzh. 2 (1963) 78; Ikonnikov, Opred. rast. Pamira (Key to Plants of Pamir) (1963) 78, p.p. —*K. capillifolia* auct. non Clarke: Kük. in Pflanzenreich [IV, 20], 38 (1909) 36, p.p.; Sergievsk. in Fl. SSSR, 3 (1935) 109, p.p. —*K. bellardii* auct. non Degl.: Sergievsk., l.c. 109. p.p. —*K. filifolia* auct. non Clarke: Sergievsk. l.c. 109, p.p. quoad pl. Asiae med.

Described from Mid. Asia (Alay valley). Type in Leningrad. Plate VII, fig. 7; map 2.

In wet and marshy, often rather saline meadows, along banks of rivers and brooks and rocky slopes; typical high-altitude plant found occasionally in middle and upper parts of forest belt.

IA. Mongolia: ***Mong. Alt.*** (10 km south-east of Yusun-Bulak, central portion of nor. trail of Khan-Taishiri mountain range, steppe, 14 VII 1947—Yun.).

IIA. Junggar: ***Jung. Alt.*** (upper Khorgos, 2700 m, VIII 1878—A. Reg.; Borotala river basin below Koketau pass, 21 VII; same site, toward Kara-ungur pass, 22 VII 1909—Lipsky; between Syaed and Ven'tsyuan, on forest slope, 2000 m, No. 3416, 13 VIII; 30 km west of Ven'tsyuan, 2640 m, Nos. 2030 and 2061, 25 VIII 1957—Kuan); ***Tien Shan.***

IIIC. Pamir (Billuli river at confluence with Chumbus, along brook, 11 VI; Kenkol river, Tom-Kara area, along marshy bank of brook, 14 VI; Ulug-tuz gorge in Charlysh river basin, along bank of brook, 28 VI 1909—Divn.; Terart pass, 27 VII; Kashka-su pass, 28 VII 1913—

Knorring; Til'nen gorge, 4500–5000 m, 1 VII; between Atrakyr and Tyuzutek rivers, mossy tundra, 20 VII; Taspetlyk area, 4000–5000 m, 25 VII; Goodzhiro river, 4500–5500 m, 27 VII 1942—Serp.; King-tau mountain range, 3–4 km south-east of Kosh-Kulak settlement, in spruce thicket, 3250 m, 10 VI 1959—Yun.).

General distribution: Jung.-Tarb. (Jung. Alatau), North and Centr. Tien Shan, East. Pam. (less frequent); Mid. Asia (Pamiroalay—Alay and Turkestan mountain ranges), North Mong. (Hang.).

Note. This species, extensively distributed in Tien Shan, is found in Pamiroalay, although rather infrequently. It differs from the related Tibetan-Himalayan species *K. capillifolia* in dimensions of the perigynium and nut as well as in the colour of glumes. In *K. capilliformis*, the perigynium measures 2 (2.5)–3.2 (3.5), very rarely 4 mm long and nuts (without stylopodium) 2.5–2.7 (3) mm long; glumes are usually chestnut-brown and without white membranous margin. The perigynium in *K. capillifolia* is 4–4.5 mm long and nut (without stylopodium) 3.3–3.5 mm long and, from the data available in the literature (Kükenthal, l.c. and Clarke, l.c.), 4 mm long; glume mostly light rusty-brown with white membranous margin. Moreover, the perigynium in *K. capilliformis* is difficult to separate from the axis of the inflorescence while separation is easy in *K. capillifolia*.

N.A. Ivanova (l.c.) differentiated the two species on the basis of the form of nuts and their apexes and length-to-breadth ratio. In our view, however, they differ only in the latter characteristic. The nuts, obovate in both species, are somewhat more oblong in *K. capillifolia*.

K. capilliformis is closely related to *K. ovczinnikovii*, which appears to hold an intermediate position between the former species and *K. capillifolia*.

4. **K. filifolia** (Turcz.) Clarke in J. Linn. Soc. (London) Bot. 20 (1883) 381; Sergievsk. in Fl. SSSR, 3 (1935) 109, p.p., quoad pl. sibir.; Ivanova in Bot. zh. 24: 5–6 (1939) 483; Grubov, Consp. fl. MNR (1955) 81. —*K. capillifolia* var. *filifolia* (Turcz.) Kük. in Ofver. Finska Vet. Soc. Förhandl. 45: 8 (1902–1903) 1; idem, Pflanzenreich [IV, 20], 38 (1909) 36, p.p. quoad pl. sibir. —*K. simpliciuscula* auct. non Mackenz.: Egorova in Bot. mater. Gerb. Bot. inst. AN SSSR, 19 (1959) 79, p.p. —*Elyna filifolia* Turcz. in Bull. Soc. natur. Moscou, 28: 1 (1855) 353.

Described from East. Siberia (Transbaikal). Type in Leningrad. Plate VII, fig. 2.

In meadows and on steppe slopes and rocks; in subalpine and forest belts.

IA. Mongolia: ***Khobd.*** (south. peak of Kharkhira, 24 VII 1879—Pot.); ***Mong. Alt.*** (east. slope of Khan-Taishiri mountain range, 2600–2700 m, 10 VIII: same site, nor. slope of ridge, 21 IX 1945, Leont'ev; Khan-Tairshiri mountain range 15 km south-east of Yusun-Bulak, sedge-fescue meadow, forest fringe, 1 IX 1948—Grub.). ***Cent. Khalkha, East. Mong., Gobi-Alt.*** (south. slope of Ikhe-Bogdo mountain range, Narin-Khurimt ravine, middle hill belt, subalpine steppe, 5 IX 1943—Yun.; same site, Narin-Khurimt gorge, on rocks, 2900 m, 28 VII 1948—Grub.; north slope of Ikhe-Bogdo mountain range, midportion of Tsagan-Burgas ravine, subalpine steppe, 15 IX 1943—Yun.).

IIIA. Qinghai: ***Nanshan*** (South.-Tetungsk mountain range, in maple forest, 2250 m, 9 VIII 1880—Przew.); ***Amdo*** (30 km east of Gunkhe, hilly forest steppe, willow thickets, 3300 m, 7 VIII 1959—Petr.).

General distribution: Arct. (Asiat.), West. Sib. (Altay), East. Sib. (excluding Sayan mountain range), Nor. Mong., China (Nor., Nor.-west.—Gansu province).

5. **K. ovczinnikovii** Egor. sp. nov. —*K. capillifolia* ssp. *pamirica* Ovcz. in Fl. Tadzh. 2 (1963) 78, descr. ross. —*K. capillifolia* auct. non Clarke: Sergievsk. in Fl. SSSR, 3 (1935) 109, p.p. —*K. capilliformis* auct. non Ivanova: Ikonnikov, Opred. rast. Pamira (Key to Plants of Pamir) (1963) 78, p.p. —Planta perennis, (15) 25–40 cm alt., dense caespitosa. Caules ca. 0.8–1.2 mm in diam., leves vel paulo scabri, teretiusculi, firmi, basi vaginis numerosis aphyllis atro-fuscis coriaceis circumadti. Folia setiformi-convoluta, 0.8–1 mm lt., rigida. Inflorescentia 1.5–3.5 cm lg., lucido-fusca, plus minusve oblongo-cilindracea. Squamae ad 6 mm lg., ferrugineo-cinnamomeae vel lucido-ferrugineo-cinnamomeae, apice albo-membranaceae. Perigynia (3.8) 4–5 mm lg., ab axe spiculae facile secedentia. Stigmata 3, raro 2. Achenium 2.5–2.7 (3) mm lg., obovatum, in rostrum brevis sensim angustati.

Typus: USSR, Tadzhikistania, Pamir, in valle fl. Pschart occidentalis, 5 km supra ostium Dzhan-Kainda, ad glaream, 3660 m s.m., 5 VII 1958, N.N. Tzvelev. In Herb. Inst. Bot. Acad. Sci. USSR (Leningrad) conservatur.

Affinitas: Ab speciebus proximis *K. capilliformis* Ivanova et *K. capillifolia* (Decne.) Clarke differt: a primo—perigyniis longioribus (3.8) 4–5 mm lg. [nec (2) 2.5–3.2 (3.5) mm lg.], ab axe spiculae facile (nec difficulter) secedentibus, et squamis ferrugineo-cinnamomeis vel lucido-ferrugineo-cinnamomeis (nec plerumque castaneis), a secundo—acheniis minoribus 2.5–2.7 (3) mm lg. (nec 3.3–3.5 mm lg.).

Map 2.

In wet and marshy meadows, along banks of rivers and rocky slopes; in alpine belt.

IIIC. Pamir (Tagdumbash-Pamir, Sarykol mountain range, Pistan gorge, on rocky sites, 3200 m, 15 VII 1901—Alexeenko; 25 km south of Bulunkul', hummocky saline meadow, 12 VI 1959—Yun.; along side of Bulunkul'-Tashkurgan road, on a terrace, 3800 m, No. 00285, 12 VI; in Tashkurgan, in river valley, No. 00301, 13 VI 1959—Lee et al.).

General distribution: East. Pam.; Mid Asia (Pamiroalay).

Note. Plants described here as *K. ovczinnikovii* were classified by P.N. Ovczinnikov as subspecies of *K. capillifolia* (Decne.) Clarke. Unfortunately, however, P.N. Ovczinnikov did not provide a Latin diagnosis nor did he indicate the type, without which the taxon established by him cannot be regarded as acceptable.

6. **K. smirnovii** Ivanova in Bot. zh. 24: 5–6 (1939) 480; Grubov, Consp. fl. MNR (1955) 81; Fl. Kazakhst. 2 (1958) 35; Egorova in Bot. mater. Gerb. Bot. inst. AN SSSR, 19 (1959) 79. —*K. schoenoides* auct. non Steud.: Kük. in Pflanzenreich [IV, 20], 38 (1909) 35, p.p.; Krylov, Fl. Zap. Sib. 3 (1929) 412, p.p.; Sergievsk. in Fl. SSSR, 3 (1935) 106, p.p.

Described from West. Siberia (Altay). Type in Leningrad. Plate VII, fig. 9; map 3.

In meadows and on rubble placers in alpine belt.

IA. Mongolia: ***Khobd.*** (Tszusylan river, alpine meadow, 11 VII 1879—Pot.), ***Mong. Alt.*** (Tolbo-Kungei-Alatau mountain range, 3200 m, Kobresia meadow and alpine steppe, 5 VIII 1945—Yun.; Khan-Taishiri mountain range, nor. slope, near ridge, 21 IX 1945—Leont'ev;

Adzhi-Bogdo mountain range, upper Ara-Tszuslangin-Gola, alpine meadow, 8 VII; same site, Ara-Tszuslan—Ikhe-Gola watershed region, rubble-rocky placers in high-altitude belt, 7 VIII; same site, upper Indertiin-Gol, hill steppe, 6 VIII; same site, Indertiin-Gol—Tumurtu-Gol pass, alpine meadow, 25 VII; Bus-Khairkhan mountain range, midportion, ravine, 17 VII; Bulugun river, Kharagaitu-Khutul' pass, alpine meadow and rubble placer with snow patches, 24 VII; same site, upper course of Ketsu-Sairin-Gol river, alpine meadow, 26 VII 1947—Yun.).

IIA. Junggar: ***Alt. Region.*** (Qinhe district, Kun'tai-Chzhunkhaitsza, on slope, 2500 m, No. 1152, 5 VIII 1956—Ching); ***Tarb.*** (Saur mountain range, south. slope in valley of Khobuk river, Karagaitu river gorge, kobresia meadow in alpine belt, 23 VI 1957—Yun.).

General distribution: Jung.-Tarb.; West. Sib. (Altay): North Mong. (Hang.).

Note. *K. smirnovii* exhibits closest affinity to *K. schoenoides* (C.A. Mey.) Steud. and is widespread in the Caucasus and Asia Minor. These two species, which essentially represent a subspecies of *K. schoenoides* s.l., differ only in dimensions of nuts: in *K. schoenoides*, they measure (without stylopodium) (3) 3.3–3.5 mm long and 1.5–1.7 mm broad; in *K. smirnovii*, they are 2.5–3 (more frequently 2.7) mm long and 1.4–1.6 mm broad. Moreover, the inflorescence is slightly smaller in *K. smirnovii* than in *K. schoenoides.*

K. smirnovii is also closer to *K. pamiroalacia* Ivanova, the only difference being non-lustrous nuts in the former.

7. **K. tibetica** Maxim. in Bull. Ac. Sci. Pétersb. 29 (1883) 219; Ivanova in Bot. zh. 24: 5–6 (1939) 483. —*K. capillifolia* var. *tibetica* (maxim.) Kük. in Pflanzenreich [IV, 20], 38 (1909) 36; Hao in Englers Bot. Jahrb. 68 (1938) 585. —*K. schoenoides* auct. non Steud.: Kük., l.c. 35, p.p.; ?Duthie in Alcock, Rep. Pamir. Comiss. (1898) 27; Deasy in Tibet a. Chin. Turk. (1901) 398; Hemsley, Fl. Tibet (1902) 200; Strachey, Catal. (1906) 200; ?Pampanini, Fl. Carac. (1930) 82; Walker in Contribs. U.S. Nat. Herb. 28 (1941) 600.

Described from Tibet (Nanshan). Type in Leningrad. Plate IV, fig. 4; map 3.

In swamps and marshy, sometimes rather saline meadows along banks of rivers and lakes in high-altitude belt; forms tufts.

IB. Kashgar: ***West.*** (upper course of Tiznaf river, Kyude settlement, 161 km south of Kargalyk on highway to Tibet, short-grass meadow along brook, 1 VII 1959—Yun.; same site, 3030 m, No. 425, 31 V 1959—Lee et al.); ***South.*** (Cherchen district 30 km south-east of Achan in high-altitude belt, 4009 m, No. 9431, 4 VI 1959—Lee et al.).

IIIA. Qinghai: ***Nanshan*** (in swamps along Kuku-Nor lake, common, 3600 m, 13 VII 1880—Przew., typus!; Solomo locality in Barduna valley, 16 V; along Tashitu river, 25 V; hills between Khsan and Tashitu, 25 V 1886—Pot.; Sharagol'dzhin river, Baga-Bulak area, along swamp, hummocks, 4500 m, 15 VI 1894—Rob.; Mon'yuan', glacial moraine at sources of Ganshiga river, Peishikhe tributary, 3900–4300 m, 18 VIII 1958—Dolgushin; "La Chi Tzu Shan, in tufts, on an exposed moist steppe, No. 692, 1923, R.C. Ching"—Walker, l.c.); ***Amdo*** ("Kokonor, Tsi-gi-gan-ba, in der Nähe des Klosters Ta-schiu-sze, zwischen 3600 m und 3900 m, sehr häufig anzutreffen, nos. 1014 and 1015, 26 VIII 1930"—Hao, l.c.).

IIIB. Tibet: ***Chang Tang*** (left bank of Karakash river, upper course of Khotan-Darya river, 10–15 km west of Shakhidulla, saline meadow, 3900 m, 3 VI; Raskem-Darya river valley, Yarkenda river upper course, near Mazar settlement, saline meadow, 4 VI 1959—Yun.; "Tibet, 34°51′ N lat., 82°16′ E long., a small patch only, on the bare hill-side; on dry gravel near the bed of a dry mountain stream, 4890 m, No. 839, 19 VII 1896, Deasy and Pike"—Deasy, l.c.; Hemsley, l.c.); ***Weitzan*** (Burkhan-Budda mountain range, Nomokhun river valley, in a swamp, 23 V 1884—Przew.; same site, along river bank, 4200 m, 22 V 1900—Lad.; Huang He river basin, hills along Bychu river, marshy bank of river, 15 VI 1884—Przew.; basin of Yantszytszyan river, Yugin-do area, in hummocky marshes, 3900 m, 13 V 1901—Lad.); ***South.*** (Kambajong, No. 36, 8–10 IV 1903—F.E. Younghusband).

IIIC. Pamir (near Bulunkul'-Tashkurgan road, No. 00287, 12 VI 1959—Lee et al.; "Pamir region, 3900–4200 m, No. 17766, Alcock"—Duthie, l.c.).

General distribution: Near East (Hindukush-Chitral), China (North-west.—Gansu province, South-west.—north Sichuan), Himalayas.

Note. Species extremely typical of high-altitude marshy sites in the southern regions of Central Asia. *K. tibetica* is very close to *K. pamiroalaica* Ivanova, inhabiting Pamiroalay, East. Pamir, West. and occasionally Cent. Tien Shan as well as the Himalayas. The inflorescence in *K. tibetica*, however, is a simple spike, very similar externally to the corresponding one in plants of genus *Eleocharis*, but in *K. pamiroalaica*, compound and spicate. Moreover, *K. tibetica* usually has a few more light-coloured glumes than *K. pamiroalaica*.

K. tibetica varies in general habit, sometimes exhibiting massive growth (specimens from Karakash river basin). *K. schoenoides*, in which the aforesaid authors identified the Cent. Asian and Himalayan plants, is widespread in the Caucasus and Asia Minor. It differs from *K. tibetica* in compound inflorescence and much larger non-lustrous nuts.

Section Psammostachys Ivanova

8. **K. robusta** Maxim. in Bull. Ac. Sci. Pétersb. 29 (1883) 218; Kük. in Pflanzenreich [IV, 20], 38 (1909) 36; Ivanova in Bot. zh. 24: 5–6 (1939) 488. —*K. sargentiana* Hemsl. in J. Linn. Soc. (London) Bot. 30 (1894) 139; ejusd. Fl. Tibet (1902) 200; Deasy, in Tibet a. Chin. Turk. (1901) 398. —*K. robusta* var. *sargentiana* (Hemsl.) Kük. l.c. 36. —**Ic.:** Pflanzenreich [IV, 20], 38 fig. 7.

Described from Tibet (Nanshan). Type in Leningrad. Plate VII, fig. 1; map 1.

On sand along banks of rivers and lakes in high-altitude belt.

IC. Qaidam: ***Mountain.*** (nor. slope of Ritter mountain range, sands, 3300 m, 22 VI 1894—Rob.).

IIIA. Qinghai: ***Nanshan*** (near Kuku-Nor lake, No., 473, 13 VII 1880—Przew., typus!).

IIIB. Tibet: ***Chang Tang*** (north slope of Przewalsky mountain range, in foothills, on moist sand, 20 VIII 1890—Rob.; "Tibet, lat. N 34°55'38", long. E 82°15'20", Camp 17, 4890 m, 19 VII 1896"—Deasy, l.c.); ***Weitzan*** (valley of Alyk-Norin-Gol river, dry sandy banks 3 VI 1900—Lad.; "Tibet, hill slope two miles N. of Murus river, head-waters of Yangtsekiang, at 4425 m, lat. N 33°53', long. E 91°31', sandy soil, some clay, 21 VI 1892, Rockhill"—Hemsley, l.c. 1894, 1902).

General distribution: North Mong. (Hang.).

Note. Nearly exclusive Centr. Asian species. Varies in general habit and size of inflorescence. Specimens with smaller inflorescence were regarded by Kükenthal (l.c.) and N.A. Ivanova (l.c.) as variant of *K. robusta*—*K. robusta* var. *sargentiana*. In our view, these plants do not merit recognition as a distinct taxon.

Subgenus **Kobresia**

Section Kobresia

9. **K. humilis** (C.A. Mey. ex Trautv.) Serg. in Fl. SSSR, 3 (1935) 111, excl. pl. caucas.; Ivanova in Bot. zh. 24: 5–6 (1939) 495; Fl. Kirgiz. 2 (1950)

263; Fl. Tadzh. 2 (1963) 80. —*K. royleana* var. *humilis* (C.A. Mey. ex Trautv.) Kük. in Pflanzenreich [IV, 20], 38 (1909) 46, p.p. —*K. simpliciuscula* auct. non Mack.: Sergievsk. l.c. 110. —*Elyna humilis* C.A. Mey. ex Trautv. in Acta Horti Petrop. 1: 1 (1871) 21. —**Ic.:** Fl. SSSR, 3, Plate IX, fig. 2.

Described from Jung. Alatau. Type in Leningrad. Plate VII, fig. 3.

In wet meadows, along river banks, near springs and on steppe-covered slopes; mid-montane belt, less often high-altitude species, from 2500–3200 m above sea level.

IA. Mongolia: ***Khobd., Mong. Alt., Val. Lakes, Gobi-Alt.***

IB. Kashgar: ***Nor.*** (Aksu district, west of Oi-Terek, alpine meadow, 3200 m, No. 8301, 10 IX 1958—Lee and Chu; Shaya, 2740 and 2620 m, Nos. 9989 and 10035; 25 and 26 VII; Kucha, 2620 m, No. 10048, 26 VII 1959—Lee et al.).

IIA. Junggar: ***Jung. Alt.*** (Toli district, on a slope, 2450 m, Nos. 1190 and 1212, 6 VIII; 20 km south of Ven'tsyuan', 2810 m, No. 1485, 14 VIII; Ven'tsyuan', 10 km east of Taldy, in a forest, 2300 m, No. 2094, 26 VIII 1957—Kuan); ***Tien Shan.***

General distribution: Jung.-Tarb., Nor. and Cent. Tien Shan; Mid. Asia (West. Tien Shan, Pamiroalay); Nor. Mong. (Hang.).

10. **K. kansuensis** Kük. in Notizbl. Bot. Gart. Mus. Berlin, 10, 99 (1930) 881; idem, Acta Hort. Gotoburg, 5 (1930) 38; Ivanova in Bot. zh. 24: 5–6 (1939) 492.

Described from China (Gansu province). Type in Berlin.

In alpine meadows.

IIIB. Tibet: ***Weitzan*** (left bank of Dychu river [Yangtze], near Konchyunchu pass, alpine meadow, 1 VII 1884—Przew.; basin of Yantszytszyan river, along Khichu river, Nyamtso district, 4200 m, 12 VII 1900—Lad.; Burkhan-Budda mountain range, south. slope—Lad.).

General distribution: China (Nor.-west. and South-west.).

11. **K. persica** Kük. et Bornm. in Österr. Bot. Zeitschr. 47 (1897) 133; Ivanova in Bot. zh. 24: 5–6 (1939) 496; Fl. Kirgiz. 2 (1950) 264; Fl. Tadzh. 2 (1963) 81; Ikonnikov, Opred. rast. Pamira (Key to Plants of Pamir) (1963) 78. —*K. royleana* var. *humilis* (C.A. Mey. ex Trautv.) Kük. in Pflanzenreich [IV, 20], 38 (1909) 46, p.p. —*K. humilis* (C.A. Mey. ex Trautv.) Serg. in Fl. SSSR, 3 (1935) 111, p.p., quoad pl. caucas. —**Ic.:** Kük. et Bornm. l.c.; Fl. Tadzh. 2, Plate XVII, figs. 4–6.

Described from Iran. Type in Vienna. Plate VII, fig. 8.

IIIC. Pamir. (Billuli river at its confluence with Chumbus, along bank of brook, 11 VI 1909—Divn.; Kashka-su pass, 28 VII 1913—Knorring).

General distribution: East. Pam.; Caucasus, Asia Minor, Near East, Mid. Asia (Pamiroalay).

Note. This species differs from the related and externally very similar species *K. humilis* (C.A. Mey.) Serg. in a somewhat narrower inflorescence, darker coloured glumes and nuts with conical (not broadly orbicular) tip as a result of stylopodium—beak preserved in the fruit. The stylopodium in *K. humilis* is almost wholly withered.

12. **K. pusilla** Ivanova in Bot. zh. 24: 5–6 (1939) 496. —*K. royleana* var. *humilis* auct. non Kük.: ?Kük. in Acta Hort. Gotoburg, 5 (1930) 39.

Described from Tibet. Type in Leningrad.
On slopes.

IIIB. Tibet: ***Weitzan*** (Dychu river, 20 VI 1884—Przew.; Burkhan-Budda mountain range, nor. slope, Nomokhun gorge, on clayey and clayey-sandy soil, 4200 m, 22 V 1900—Lad.; Yantszytszyan river basin, Bana-Chokun village, on wet grass plots, 4500 m, 12 IV 1901—Lad., type!).

General distribution: ?China (South-west.—north Sichuan).

Note. This species is known only from the sites listed above. None of the specimens had developed nuts and hence it is difficult to consider them an independent species. Yet, *K. pusilla* differs from *K. humilis* in varying number of stigmas (specimens of V.F. Ladygin have styles with two and those of N.M. Przewalsky three stigmas). This feature, as well as the isolated range, would seem to indicate, as suggested by N.A. Ivanova, that the cited plants be treated as an independent species.

13. **K. pygmaea** (Clarke) Clarke in Hook. f. Fl. Brit. Ind. 6 (1893) 696; Kük. in Pflanzenreich, [IV, 20], 38 (1909) 39; idem, Acta Hort. Gotoburg. 5 (1930) 37, p.p. (?excl. var. *filiculmis* Kük.); Pampanini, Fl. Carac. (1930) 82; Ivanova in Bot. zh. 24: 5–6 (1939) 498. —*Hemicarex pygmaea* Clarke in J. Linn. Soc. (London) Bot. 2 (1883) 384.

Described from the Himalayas. Type in London (?K).
In meadows in high-altitude belt.

IIIB. Tibet: ***Weitzan*** (nor. slope of mountain range between Huang He and Yangtze, in a meadow, 4200 m, common, 12 VI 1884 —Przew.).

General distribution: China (south-west.—north Sichuan), Himalayas (west. and east.).

14. **K. royleana** (Nees) Boeck. in Linnaea, 39 (1875) 8; Hemsley in J. Linn. Soc. (London) Bot. 30 (1894) 124; Forbes and Hemsley, Index Fl. Sin. 3 (1905) 268; Strachey, Catal. (1906) 202; Kük. in Pflanzenreich [IV, 20], 38 (1909) 45, p.p.; Pampanini, Fl. Carac. (1930) 82; Ivanova in Bot. zh. 24: 5–6 (1939) 492. —*Trilepis royleana* Nees in Edinb. New Phil. J. 17 (1834) 267 and in Wight, Contribs. Bot. India (1834) 119.

Described from the Himalayas (Nepal). Type in London (?BM).

IIIB. Tibet: ***Chang Tang*** ("Tibet, Kuen-lun, plains about 5100 m, H.P. Picot"—Hemsley, l.c.: "Lingzi-tang ed Akhsáe-cin, ca. 5175 m, Picot"—Pampanini, l.c.); ***South.*** (Gyangtse, No. 25, VII–IX 1904—H.J. Walton).

General distribution: China (South-west.), Himalayas.

Note. Quite possibly, plants from Chang Tang belong to *K. stenocarpa* (see note under *K. stenocarpa*).

15. **K. simpliciuscula** (Wahl.) Mack. in Bull. Torr. Bot. Club, 50 (1923) 349; Sergievsk. in Fl. SSSR, 3 (1935) 110, p.p.; Ivanova in Bot. zh. 24: 5–6 (1939) 495; Grubov, Consp. fl. MNR (1955) 81; Egorova in Bot. mater. Gerb. Bot. inst. AN SSSR, 19 (1959) 79, p.p. —*Carex simpliciuscula* Wahl. in Svensk. Vet. Ac. Handl. 24 (1803) 141. —*K. caricina* Willd. Sp. pl. 4 (1805) 206; Kük. in Pflanzenreich [IV, 20], 38 (1909) 45. —*K. bipartita* Dalla Torre, Anleit Beob. Alpenreisen (1882) 330; Krylov, Fl. Zap. Sib. 3 (1929) 415. **Ic.:** Fl. SSSR, 3. Plate IX, fig. 6.

Described from England. Type in Stockholm. Plate VII, fig. 5.

In meadow steppes, dry meadows and on steppe and rock slopes.

IA. Mongolia: ***Mong. Alt.*** (Khan-Taishiri mountain range, north slope, lower hill belt, meadow steppes along ravine floors, 11 VII 1945—Yun.; midportion of north trail of Khan-Taishiri mountain range, herb-chee grass steppe, 14 VII 1947—Yun.); ***Gobi-Alt.*** (south. slope of Ikhe-Bogdo mountain range, midportion of Narin-Khurimt ravine, north rocky slope, 28 VI 1945—Yun.).

IIA. Junggar: ***Jung. Gobi.***

General distribution: Arct., Europe (West. Europ. hills, Nor. and Cent. Urals), West. Sib. (Altay), East. Sib., Far East (Okhotsk coast—Ayan), North Mong. (Fore Hubs., Hent.), North Amer.

16. **K. stenocarpa** (Kar. et Kir.) Steud. Syn. Cyp. (1855) 246; Ivanova in Bot. zh. 24: 5–6 (1938) 492; Fl. Kirgiz. 2 (1950) 263; Fl. Kazakhst. 2 (1958) 37; Fl. Tadzh. 2 (1963) 79; Ikonnikov, Opred. rast. Pamira (Key to Plants of Pamir) (1963) 78. —*Elyna stenocarpa* Kar. et Kir. in Bull. Soc. natur. Moscou, 15 (1842) 526. —*K. paniculata* Meinsh. in Acta Horti. Petrop. 18, 3 (1901) 279; Sergievsk. in Fl. SSSR, 3 (1935) 111. —*K. royleana* var. *paniculata* (Meinsh.) Kük. in Pflanzenreich [IV, 20], 38 (1909) 46. —*K. royleana* var. *kokanica* (Rgl.) Kük. l.c. 46. —*K. royleana* auct. non Boeck.: Kük. l.c. 45, p.p.; Sergievsk. l.c. 110; Duthie in Alcock, Rep. Pamir Commis. (1898) 27; Persson in Bot. Notis. (1938) 276. —**Ic.:** Fl. Tadzh. 2, Plate XVII, figs. 1–3.

Described from Junggar Alatau. Type in Leningrad. Plate VII, fig. 4; map 4.

In wet and marshy Carex-Kobresia meadows, marshy hummocky short-grass meadows along banks of rivers and brooks, on pebble beds, less frequently on rubble and steppe slopes, predominantly in high-altitude belt.

IIA. Junggar: ***Jung. Alt.*** (Kasan pass, 2700–3300 m, 11 VIII 1878—A. Reg.; 20 km south of Ven'tsyuan', No. 1559, 14 VIII 1959—Lee et al.); ***Tien Shan.***

IIIA. Qinghai: Nanshan (north slope of Humboldt mountain range, near Ulan-Bulak area, along brook, 3600 to 3900 m, 18 VI 1904—Rob.).

IIIB. Tibet: ***Chang Tang*** (north slope of Przewalsky mountain range, on sand, 4200 m, 22 VIII 1890—Rob.; upper Khotan-Darya river, 10–12 km west of Shakhidulla, short-grass meadow, 3900 m, 3 VI; same site, 25–30 km east of Kirgiz-Dzhangil pass, on road to Shakhidulla, saline meadow, 3 V 1959—Yun.).

IIIC. Pamir.

General distribution: Jung.-Tarb., North and Centr. Tien Shan, East. Pamir; Mid. Asia (West. Tien Shan, Pamiroalay), East. Sib. (East. Sayan mountain range).

Note. *K. stenocarpa* is closer to *K. royleana*. According to N.A. Ivanova (l.c.), the two species differ in that the nuts in *K. stenocarpa* are rather oblong with parallel sides and the perigynium exceeds the nut by 1.5–2 times, while in *K. royleana*, the nuts are narrowly obovate and the perigynium only slightly longer than the nut. In the specimens of *K. stenocarpa* studied by us from Nor. and Cent. Tien Shan, Junggar and Pamir, contrary to N.A. Ivanova's observation, the perigynium is almost equal to the mature nut with the beak (stylopodium) only very insignificantly exceeding it. The Herbarium of the Botanical Institute of the Academy of Sciences, USSR, has four specimens of *K. royleana* from the Himalayas, none of them with a developed nut. Ripe fruits are present in one herbarium specimen from Tibet (Gyangtse) which, in all probability, belongs to *K. royleana*. Its nuts differ from those of *K. stenocarpa* in shape (they are narrowly obovate) as well as in very short (0.4 mm long) beaks (beak of *K. stenocarpa* 0.7–0.9 mm long). Moreover, the margin of the perigynium in the mature nut of the Tibetan plant is closed only in the lower portion while this closure almost reaches the tip in *K. stenocarpa*. The extent to which these differences persist can be resolved by studying more specimens of *K. royleana*. Plants belonging to *K. stenocarpa* or *K. royleana* from Qinghai and Tibet, not possessing a developed nuts, have been tentatively placed by us among *K. stenocarpa*.

Section Chlorostachys Ivanova

17. **K. prattii** Clarke in J. Linn. Soc. (London) Bot. 36 (1905) 268; Forbes and Hemsley, Index Fl. Sin. 3 (1905) 268; K‥ik. in Pflanzenreich [IV, 20], 38 (1909) 43; Rehder in J. Arnold Arb. 14 (1933) 4; Ivanova in Bot. zh. 24: 5–6 (1939), 500. —*K. harrysmithii* Kük. in Acta Hort. Gotoburg. 5 (1930) 37.

Described from China (Sichuan). Type in London (K).

IIB. Tibet: ***Weitzan*** (Yantszytszyan river basin, Guda-choki area, Gonchu river, along hill slopes, 3840 m, 1 V 1901—Lad.).

General distribution: China (North, North-west., South-west.—north Sichuan).

Carex L.

Sp. pl. (1753) 972; idem, gen. pl., ed. 5 (1754) 420.

1. Inflorescence with single-spikelet. Spikelet androgynous ....2.
\+ Inflorescence with two or more spikelets ....5.
2. Utricles almost subulate. Glumes caducous (not seen in developed utricles)...4. **C. microglochin** Wahl.
\+ Utricles ovate, broadly ovate or oblong-ovate, abruptly contracted into short beak. Glumes not caducous ....3.
3. Plant with slender, long creeping rhizome covered, like stem base, with purple-coloured glumes. Utricles ovate or broadly ovate, 2.5–3.5 mm long, lustrous, blackish-brown; mature ones declinate from axis of spikelet. ....2. **C. obtusata** Liljebl.
\+ Plants with long or short creeping rhizome covered with light brown or chestnut-brown glumes. Utricles obovate or oblong-obovate, non-lustrous, rusty-borwn; mature ones appressed to axis of spikelet ....4.
4. Utricles convexly trigonous, elliptical or obovate, (2.5) 2.8–3 mm long, with rachilla at base; glumes rusty light brown. Spikelets thickened, 5–6 mm broad, 1.2–2 cm long, with well-developed staminate part consisting of several staminate flowers (not less than 30). Leaves erect or falcate but not sinuate............................. ....1. **C. argunensis** Turcz. ex Trev.
\+ Utricles trigonous, 3.2–4 mm long, oblong-ovate or oblong-obovate, usually without rachilla at base; glumes chestnut-coloured. Spikelets very narrow, 3–4 mm broad, 1–1.5 (2) cm long;

staminate flowers fewer (less than 20). Leaves sinuate towards upper part, with withered tips.............3. **C. rupestris** Bell. ex All.

5 (1). Stigmas two; fruits biconvex..6.
+ Stigmas three; fruits trigonous...30.
6. All spikelets in inflorescence almost similar in form and size, bisexual, less frequently unisexual; in latter case terminal spikelet not staminate...7.
+ Spikelets in inflorescence of two types: upper one (1–3) very narrow and staminate; lower ones thickened and pistillate (less frequently, pistillate spikelets with a few staminate flowers at tip).............25.
7. Spikelets in inflorescence or all of hem androgynous (i.e., with staminate flowers in upper part of spikelet and pistillate in lower), sometimes mixed with staminate spikelets (in midportion of inflorescence) or unisexual (lower and upper ones—pistillate and central ones—staminate), sometimes mixed with androgynous ones...8.
+ Spikelets in inflorescence gynaecandrous (i.e., with pistillate flowers in upper and staminate in lower part of spikelet), sometimes mixed with staminate flowers...22.
8. Rhizome not creeping; plants densely caespitose..........................9.
+ Rhizome creeping; plants lax, less often densely caespitose......10.
9. Inflorescence with several spikelets, brown. Stems thickened, 2–4 mm in diameter, flatten under pressure. Utricles reddish or rusty-brown, with thickened veins in front......................5. **C. vulpina** L.
+ Inflorescence with 5–10 spikelets, green. Stems slender, 1–1.5 mm in diameter; do not flatten. Utricles green or brownish-green, without veins or with very indistinct veins.................................... ....6. **C. polyphylla** Kar. et Kir.

10 (8). Utricles with dense and short pubescence towards upper part..... ....*__C. pallida__ C.A. Mey. p. 56.
+ Utricles glabrous...11.
11. Rhizome very long, slender, 0.8–1.5 (2) mm in diameter; at ends with fascicles of two or more shoots; less frequently shoots arise singly. Plants usually of dry habitats (sand, silt)........................12.
+ Rhizome often thickened (2.5–5 mm in diameter), with shoots separated and arising singly; less often 2 or 3 shoots together. Plant of wet and marshy habitats..15.
12. Utricles coriaceous, easily separated from axis of spikelet.........13.
+ Utricles membranous, not easily separated from axis of spikelet, often together with adjoining flowers..14.
13. Sheath surrounding shoot base mostly grey, fibrous. Inflorescence 0.7–1.3 cm long; glumes brown, with narrow white membranous margins; colourless on fading only at end of growth cycle. Utricles

broadly ovate, less often ovate, 2.5–3 (3.2) mm long, abruptly contracted into beak, usually without veins. Roots usually grey............ ....12. **C. duriuscula** C.A. Mey.

\+ Sheath surrounding base of shoots usually brown, slightly fibrous or not. Inflorescence 1.5–2.5 cm long; glumes brown, with broad white membranous margins, sometimes almost completely white. Utricles usually ovate, less often broadly ovate, (3) 4–4.5 (5) mm long, gradually contracted into beak, with veins. Roots usually light brick-red...............................14. **C. stenophylloides** V. Krecz.

14 (12) Utricles strongly inflated, up to 2 cm long, with several slender veins and smooth beak. Base of stems surrounded by loosely arranged long and broad light brown sheaths, somewhat fibrous towards upper part. Psammophyte................13. **C. physodes** M.B.

\+ Utricles not more than 5.5 mm long, not inflated, without or with indistinct veins at base, with sparsely scabrous, less often smooth beak. Sheath at base of stems short, blackish-brown or chestnut-brown, fibrous, densely surrounds shoots. Plant of clayey semideserts and desert-steppe montane belt ....*C. **pachystylis** J. Gay., p. 58.

15 (11) Rhizome with thick, compact ligneous bark which does not peel or collapse on desiccation..16.

\+ Rhizome with thin loose bark which peels off and usually collapses on desiccation[1] ....18.

16\. Upper spikelet in inflorescence pistillate, middle ones staminate or sometimes androgynous; lower ones pistillate, sometimes mixed with staminate ones; pistillate spikelets 0.8–1.5 cm long. Utricles slightly flattened, with thin veins................8. **C. lithophila** Turcz.

\+ Upper spikelets or all of them in inflorescence androgynous, middle ones sometimes also staminate; spikelets 4–6 mm long. Utricles not flattened, with thickened veins..17.

17\. Inflorescence narrow, rather oblong; glumes usually ovate, obtuse, brown, shorter than utricles. Utricles oblong-ovate or ovate-lanceolate, with elongated beak, pale between veins; veins in mature utricles ribbed, very prominent........................7. **C. curaica** Kunth.

\+ Inflorescence broad, ovate; glumes usually oblong-ovate, acute or acuminate, less often obtuse, light brown, with rather broad, white membranous margins equal to utricles. Latter ovate or ovate-elliptical, with short beak, straw-yellow and brownish between veins; veins in mature utricles rather smoothened.. ....9. **C. pycnostachya** Kar. et Kir.

---

[1]The structure of the rhizome can be easily determined by macroscopic slicing with a blade.

18 (15) Stems indistinctly trigonous, subcylindrical, smooth or only slightly scabrous under inflorescence. Glumes ovate. Utricles slightly but perceptibly inflated and biconvex, generally ovate, with smooth or slightly scabrous beak; utricles not enveloping fruit entirely..19.

\+ Stems trigonous, scabrous. Glumes oblong-ovate; utricles planoconvex, barely inflated, usually oblong-ovate, usually with scabrous margin and beak; utricles enveloping fruit almost completely ....21.

19. Stems thickened, 1.8–2.5 mm in diameter; leaves flat or folded, 2–4 mm broad. Inflorescence 1.3–2 cm long, 0.8–1.7 cm broad; utricles 3.5–4.5 mm long; glumes dark chestnut-brown...16. **C. pseudofoetida** Kük.

\+ Stems slender, 0.8–1.3 mm in diameter; leaves folded or convoluted like a bristle, 1–2 mm broad. Inflorescence 0.5–1 cm long, not more than 1 cm broad; utricles 3–3.5 (4) mm long; glumes rust coloured ....20.

20. Stems 5–20 cm long. Inflorescence orbicular, densely capitate; glumes ovate, generally obtuse, dark rust coloured..19. **C. sajanensis** V. Krecz.

\+ Stems 15–45 cm long. Inflorescence oblong-ovate, rather lax; glumes oblong-ovate, acute, rust coloured..17. **C. reptabunda** (Trautv.) V. Krecz.

21(18) Utricles usually without veins..................15. **C. enervis** C.A. Mey.

\+ Utricles with veins ....18. **C. roborowskii** V. Krecz.

22 (7) Rhizome creeping. Utricles with narrow wing along margin in upper portion. Plants of dry habitats ....23.

\+ Rhizome shortened, not creeping. Utricles not winged. Plants of wet and marshy habitats ....24.

23. Inflorescence with (3) 4 or 5 (6) spikelets; all spikelets gynaecandrous. Utricles ovate, (3) 3.5–3.7 (4) mm long, almost coriaceous, mature ones erect, not curved ....11. **C. praecox** Schreb.

\+ Inflorescence with 5–8 spikelets; central spikelets sometimes staminate. Utricles ovate-lanceolate, (4) 4.5–5 mm long, thin-membranous; mature ones declinate from axis of spikelet and curved
....10. **C. diplasiocarpa** V. Krecz.

24. Inflorescence oblong, 2.5–5 cm long, with 4–7 spikelets, usually interrupted, pale green. Utricles green or brownish-green, with short beak slightly emarginate in front.............20. **C. canescens** L.

\+ Inflorescence capitate, 1–2 cm long, with 3 or 4 spikelets, chestnut-brown or rusty-brown. Utricles light rust, with elongated beak deeply laciniate like slit in front.......................21. **C. tripartita** All.

25 (6) Utricles with veins. Lower bract leaf usually longer than inflorescence ....26.

+ Utricles without veins. Lower bract leaf usually shorter than inflorescence ....27.

26. Plant with rather long creeping rhizome; not forming tufts. Utricles biconvex, slightly inflated; glumes generally longer than utricles ....25. **C. fuscovaginata** Kük.

+ Plant with shortened non-creeping rhizome; forming rather larger tufts. Utricles planoconvex, not inflated; glumes shorter than utricles ....22. **C. appendiculata** (Trautv. et C.A. Mey.) Kük.

27. Plants without creeping rhizome, forming tufts. Shoot base surrounded by scaly sheaths. Staminate spikelets single.................28.

+ Plants mostly with creeping rhizome; usually not tuft forming. Shoot base surrounded by sheaths extending into leaf blade. Staminate spikelets 1 or 2..29.

28. Sheath surrounding shoot base dark purple. Inflorescence usually about 3 cm long with single staminate and two, less often three pistillate spikelets. Lower bract leaf setaceous, shorter or only slightly longer than lower spikelet; very rarely almost equalling inflorescence. Glumes of pistillate spikelets ovate or oblong-ovate, blackish-brown, usually shorter than utricles. Utricles without spinules along margin in upper part, with light-coloured beak. Roots with distinct mostly greyish-white hairs............23. **C. cespitosa** L.

+ Sheath surrounding shoot base brown. Inflorescence 3–6 (10) cm long and with 1 or 2 staminate and 1 or 2 (3) pistillate spikelets. Lower bract leaf linear, equal or almost equalling inflorescence. Glumes of pistillate spikelets lanceolate, usually slightly longer than utricles. Latter mostly with stray spinules along margin in upper part and purplish-brown beak. Roots appear glabrous due to very short brown hairs visible only under binoculars ....27. **C. schmidtii** Meinsh.

29. Mature utricles conspicuously inflated (usually flattened vertically in herbarium), very closely appressed to each other and orbicular in cross-section. Glumes usually 2 or 3, sometimes 4 times narrower than utricles. Shoot base surrounded by greyish-brown (sometimes with purple tint) non-lustrous fibrous leaf sheaths. Rhizome moderately creeping; sometimes shortened; usually not every shoot in mat has distinctly manifest horizontal section (rhizome) ....26. **C. orbicularis** Boott.

+ Utricles not conspicuously inflated, not collapsed; oval in cross-section. Glumes almost as broad as utricles or slightly less than one-half narrower. Shoot base surrounded by reddish or yellowish-brown, usually lustrous entire leaf sheaths. Rhizome creeping; usually each shoot in mat has distinctly manifest horizontal section (rhizome) ....24. **C. ensifolia** (Turcz. ex Gorodk.) V. Krecz.

30 (5) Uppermost or upper 2 or 3 spikelets androgynous, rest pistillate, or all spikelets androgynous. Plants densely caespitose, without creeping rhizome ....31.

+ Uppermost spikelet (or 2 or more) staminate or gynaecandrous (very rarely pistillate spikelets with some staminate flowers on top but then rhizome creeping) ....33.

31. Leaves narrowly grooved, about 1 mm broad. All spikelets distant. Utricles 2.5–2.8 mm long, with short beak............................ ....70. **C. sedakowii** C.A. Mey. ex Meinsh.

+ Leaves flat, 1.5–2.5 mm broad. Upper spikelets crowded and lower ones distant. Utricles 4–5.3 mm long, with elongated beak....... 32.

32. Spikelets 3–4 cm long, linear-cylindrical. Utricles elliptical, 4 mm long, smooth, with many veins which gradually contract into beak. Glumes rust coloured ....61. **C. dielsiana** Kük.

+ Spikelets ovate or oblong-ovate, 1–1.5 cm long. Utricles obovate, 4.5–5.3 mm long; upper portion along margin scabrous with spinules; whole surface except base rather densely covered with tiny appressed bristle-like hairs, without veins, abruptly contracted into beak. Glumes chestnut-brown.....................57. **C. alajica** Litw.

33 (30) Uppermost spikelet gynaecandrous ....34.

+ Uppermost spikelet or some upper spikelets staminate............. 49.

34. Lower bract leaf with 1–3 cm long sheath. Utricles without veins ....35.

+ Lower bract leaf without sheath or with very short (2–3 mm long) sheath. Utricles with or without veins ....37.

35. Spikelets arranged in inflorescence such that their tips are nearly at same level; pistillate spikelets 2.5–4.5 cm long. Leaves 5–7 mm broad. Utricles flat, blackish-purple, 5–6 mm long ....66. **C. nivalis** Boott.

+ Tips of spikelets in inflorescence at different levels; pistillate spikelets not more than 1.5 cm long. Leaves 1.5–2.5 mm broad. Utricles convexly trigonous, green or greenish-brown, not more than 3 mm long ....36.

36. Mature utricles greenish-brown, 2.2–2.7 (3) mm long, ovate or oblong-ovate, lustrous ....69. **C. karoi** (Freyn) Freyn.

+ Mature utricles pale green, 1.7–2 mm long, almost broadly ovate, matte (not lustrous)...............................71. **C. selengensis** Ivanova.

37 (34) Inflorescence dense, erect, capitate or capitate-spicate, with very closely arranged spikelets (lower spikelet sometimes slightly distant); peduncle of lower spikelet usually not more than 0.5 cm long, less often up to 1 cm ....38.

+ Inflorescence rather lax, racemose, drooping or erect with declinate lateral spikelets; peduncle of lower spikelet 1–5 cm long .... 44.

38. Plants loosely or densely caespitose with rather long creeping horizontal or obliquely ascending rhizome. Inflorescence (2) 2.5–

4.5 cm long; glumes usually longer than utricles, sometimes as long. Leaves almost as long as stem .... 39.
+ Plants densely caespitose, without creeping rhizome or with very short rhizome. Inflorescence 1–2 cm long; glumes usually shorter than utricles. Leaves shorter than stem, quite often by half...... 40.
39. Stems smooth; leaves often curved, flexuose towards tip. Pistillate spikelets 1–1.5 cm long, 0.8–1 cm broad. Utricles elliptical or obovate, (4) 4.5–5 (6) mm long, abruptly contracted into bicuspid beak; less often, latter emarginate only in front; beak 0.6–1 (1.2) mm long ....40. **C. sabulosa** Turcz. ex Kunth.
+ Stems scabrous; leaves erect or somewhat curved, not flexuose. Pistillate spikelets 1.5–2.5 cm long and 0.6–0.8 cm broad. Utricles broadly obovate or broadly elliptical to suborbicular, rather abruptly contracting into entire, less often emarginate (very rarely short and bicuspid) beak 0.2–0.3 (0.5) mm long ....35. **C. melananthiformis** Litv.
40 (38) Lower bract leaf a few times longer than inflorescence. Beak of utricles smooth...33. **C. lehmannii** Drej.
+ Lower bract leaf shorter, equal to or longer than inflorescence but not more than thrice as long. Beak of utricles mostly scabrous.... 41.
41. Utricles with veins ....42.
+ Utricles without veins ....43.
42. Utricles with short but distinct rough beak...31. **C. infuscata** Nees.
+ Utricles without beak, entire or emarginate on top..........................
....*C. **pseudobicolor Boeck.**, p. 74.
43. Inflorescence mostly broadly ovate, very compact, blackish-brown. Utricles oblong-ovate or oblong-elliptical, (3) 3.5–4.5 mm long, not granulate, purplish-black in upper half. Glumes half length of utricles...36. **C. melanocephala** Turcz.
+ Inflorescence mostly ovate, less compact. Utricles mostly obovate or elliptical, 2.2–2.5 (2.8) mm long, with dense finely granulate surface, brownish-green, with very dark-coloured beak. Glumes 1/4–1/3 length of utricles ....38. **C. norvegica** Retz.
44 (37) Utricles 2–2.5 mm long. Glumes half or less than length of utricles. Inflorescence 2–2.5 (3) cm long..
....................................33. **C. lehmannii** Drej. (also see couplet 40).
+ Utricles 3–4.5 mm long. Glumes slightly shorter than utricles or as long. Inflorescence (3) 3.5–9 cm long ....45.
45. Utricles flattened-trigonous, 3.5–4.5 mm long, without or less often with indistinct veins. Lower spikelets (1.5) 1.8–3 cm long.........46.
+ Utricles trigonous, slightly inflated, 3–3.5 mm long, with 5 or 6 thin distinct veins. Lower spikelets 1.5–2 cm long.........................
....30. **C. hancockiana** Maxim.

46. Utricles diffusely spiny-ciliate along upper margin, brownish-yellow and often with brown specks; surface of utricles micropapillate in upper part. ....32. **C. kansuensis** Nelmes.
 + Utricles not ciliate along margin..47.
47. Utricles brownish-yellow. Glumes lanceolate, acute or cuspidate. Inflorescence 6–9 cm long, lax; spikelets rather lax; peduncle of lower spikelet 3–5 cm long ....28. **C. caucasica** Stev.
 + Utricles blackish-purple. Glumes ovate or oblong-ovate, acute or subacute. Inflorescence (3) 3.5–4.5 cm long, compact; spikelets dense; peduncle of lower spikelet 1–2 (3) cm long.....................48.
48. Utricles elliptical, with short (0.3–0.4 mm long) but distinct beak. ....39. **C. perfusca** V. Krecz.
 + Utricles oblong-elliptical, without or with very short beak........... ....29. **C. duthiei** Clarke.
49 (33) Utricles crenately winged along margin right from base............... 75. ....75. **C. eremopyroides** V. Krecz.
 + Utricles not winged...50.
50. Utricles strongly inflated (fruit freely placed in utricles), membranous, glabrous, with smooth beak. Pistillate spikelets cylindrical, very dense, thickened, up to 1 cm in diameter. Plants of excessively wet habitats, often very large ....51.
 + Utricles not inflated, glabrous or rather pubescent, with smooth or scabrous beak ....59.
51. Staminate spikelet one; pistillate spikelets on long peduncles, drooping. Utricles ovate-lanceolate, mature ones deflexed with elongated subulate-bicuspid beak. Plant densely caespitose, without creeping rhizome ....78. **C. pseudocyperus** L.
 + Staminate spikelets 2–4; pistillate spikelets sessile or peduncled but not drooping. Mature utricles usually not deflexed; if slightly deflexed then spherical-ovate, not ovate-lanceolate. Plants with rather long creeping rhizome. ....52.
52. Beak of utricles distinctly bicuspid, with up to 3 mm long teeth; utricles usually yellowish-green or slightly brownish. Glumes brownish...53.
 + Beak of utricles barely emarginate, without distinct teeth; utricles often dark purple or reddish-black, highly glossy ....58.
53. Glumes of pistillate spikelets long and aristate. Utricles subcoriaceous ....54.
 + Glumes of pistillate spikelets acute or subobtuse. Utricles membranous ....56.
54. Lower bract leaf with more than 1 cm long sheath. Teeth of beak in utricles subulate, 1–3 mm long; mature utricles yellowish-green, rust coloured only at base of incision of beak and along its teeth. Stems smooth; leaf sheaths rather pubescent or glabrous..........55.

+ Lower bract leaf usually without or with very short sheath. Teeth of beak in utricles soft, usually not more than 1 mm long; mature utricles brownish-green, rust coloured on top. Stems smooth or scabrous; leaf sheaths glabrous. ....82. **C. drymophila** Turcz. ex Steud.

55. Utricles ovate or oblong-ovate, 6–7 mm long, with distinct veins; beak of utricles incised to same depth in front and rear. Leaves 3–6 mm broad ....83. **C. orthostachys** C.A. Mey.

+ Utricles oblong-conical, (7) 8–10 mm long, with indistinct veins; beak of utricles incised far more in front than rear (incision in front semicircular). Leaves 2–4 mm broad ....84. **C. raddei** Kük.

56 (53) Stems in upper part sharply trigonous and strongly scabrous. Mature utricles obliquely erect, ovate or oblong-ovate, gradually contract into beak. Roots with dense yellowish hairs.....81. **C. vesicata** Meinsh.

+ Stems obtusely trigonous, smooth, less often rather scabrous but then mature utricles diverged horizontally or slightly downwards from axis of spikelet. Utricles broadly ovate; abruptly contracted into beak ....57.

57. Leaves (6) 8–15 mm broad, green, not thickened, flat. Utricles 5–6.5 mm long, turning brown early, with beak equal to third length of utricle. Pistillate spikelets usually very dense, 1–1.5 cm in diameter. Stems smooth, less often scabrous...........79. **C. rhynchophysa** C.A. Mey.

+ Leaves (2) 3–5 (8) mm broad, greyish or bluish-green, stiff, thickened, flat or sometimes folded and look like channels. Utricles 4–5 (6) mm long, yellowish-green, very rarely brownish on top, with beak about 1 mm long. Pistillate spikelets somewhat lax, usually less than 1 cm in diameter. Stems smooth.......... 80. **C. rostrata** Stokes.

58 (82) Leaves 3.5–5 mm broad. Glumes of pistillate spikelets oblong-ovate or lanceolate, usually shorter than utricles, sometimes half as long. Pistillate spikelets 2.2–2.5 (4) cm long. Plant 20–50 cm tall.............. ....76. **C. dichroa** Freyn.

+ Leaves (4) 6–8 (10) mm broad. Glumes lanceolate, equalling utricles or slightly longer. Pistillate spikelets (2.5) 3–4.5 cm long. Plant 50–100 cm tall.................................77. **C. pamirensis** Clarke.

59 (50) Plants with long, slender, creeping rhizome, with shoot clusters at ends. Utricles coriaceous, glossy, suborbicular in cross-section, usually with short beak. ....60.

+ Plants without or with creeping rhizome; in latter case, stems not clustered but distant in ones and twos; if clustered, utricles flatly trigonous...66.

60. Staminate spikelet whitish or pale brown, at same level as, or slightly below next pistillate spikelet; latter with sparse flowers, on 2.5–4 cm long peduncles. Sheath of bract leaves without leaf blade ....43. **C. alba** Scop.

+ Staminate spikelet invariably higher than next pistillate spikelet; latter usually compact. Sheath (if present) of bract leaves with distinct leaf blade ....61.

61. Staminate spikelet on 1.5–2.5 cm long peduncle (very rarely sessile if a tiny pistillate spikelet is present at its base). Leaves folded longitudinally, resembling a bristle.............48. **C. relaxa** V. Krecz.

+ Staminate spikelet sessile or on peduncle not long than 0.5 cm. Leaves flat or with margins, slightly turned on underside, less often folded longitudinally, resembling a bristle.........................62.

62. Bract leaf of lower pistillate spikelet usually with 0.5–1 cm long sheath and short, narrow or setaceous blade, less often almost reaching tip of inflorescence (sometimes sheath shorter than 0.5 cm or bract leaf scaly, but then beak of utricles 0.8–1.5 mm long). Staminate spikelets 1 or 2...63.

+ Bract leaf of lower pistillate spikelet without sheath, usually scaly, with sharp cusp. Beak of utricles not longer than 0.5 mm. Pistillate spikelet one ....65.

63. Leaves folded longitudinally, resembling a bristle, 0.5–1 mm broad. Utricles oblong-ovate, (4) 4.5–5.3 mm long, with smooth, 1–1.5 mm long elongated beak. Staminate spikelet one.................. ....45. **C. ivanoviae** Egor.

+ Leaves flat or with slightly deflexed margin, (2) 2.5–3.5 mm broad. Utricles ovate or obovate, usually rather scabrous along margins of beak and sometimes with short, setaceous hairs diffused over surface. Staminate spikelets 1 or 2 ....64.

64. Pistillate spikelets quite compact, without staminate flowers on top. Beak of utricles (0.8) 1–1.3 mm long..........49. **C. turkestanica** Rgl.

+ Pistillate spikelets rather lax, sometimes with a few staminate flowers in upper part. Beak of utricles not longer than 0.5 mm..... ....47. **C. minutiscabra** Kük.

65 (62) Glumes of staminate spikelets narrowly ovate, acute; glumes of pistillate spikelets usually acute................46. **C. korshinskyi** Kom.

+ Glumes of staminate spikelets broadly obovate with broadly rounded tips; glumes of pistillate spikelets rather acute................. ....44. **C. aridula** V. Krecz.

66 (59) Lower bract leaf with tubular 0.5–5 cm long sheath and rather developed leaf blade (less often, sheath less than 0.5 cm but then utricles hirsute) ....67.

+ Lower bract leaf without or with minute sheath. Utricles glabrous ....91.

67. Utricles flat or flattened-trigonous (sometimes only beak distinctly flattened), often (like glumes) purplish-black or reddish-brown. Fruit on fairly long slender peduncle. Staminate spikelets 1–3....... ....68.

+ Utricles obtuse- or orbicular-trigonous, with cylindrical beak, generally green, yellowish or rusty-brown; glumes brownish or greenish. Fruit sessile. Staminate spikelet one .... 76.

68. Utricles flat, with small tubercle fruit which is many times shorter and narrower than utricles; latter without veins, with smooth surface; beak smooth or with few spines along margins. Pistillate spikelet on elongated slender peduncle, drooping .... 69.

+ Utricles flattened-trigonous but not flat or trigonous; fruit not more than 2 (3) times shorter and narrower than utricles; utricles usually with veins, often covered with minute appressed setaceous hairs in upper part on both sides or only on one side; scabrous or spiny-scabrous along margins in upper part (including beak). Pistillate spikelets drooping or erect .... 72.

69. Leaves about 1 mm broad. Utricles 2 mm long. Plants about 2 cm tall. .... 65. **C. montis-everesti** Kük.

+ Leaves 3–8 mm broad. Utricles 4.5–7 mm long. Plants significantly larger .... 70.

70. Pistillate spikelets usually light brown, less frequently dark brown imparted by colour of glumes and utricles; staminate spikelet light or pale brown. Glumes of pistillate spikelets ovate or oblong-ovate; utricles generally very light coloured along margin, micropapillate in upper part (under binoculars!) .... 60. **C. coriophora** Fisch. et Mey. ex Kunth.

+ Pistillate spikelets blackish-purple due to almost fully coloured glumes and utricles; sometimes spikelets ash-coloured due to utricles with rather distinct whitish patches; glumes then lanceolate; staminate spikelet also blackish-purple .... 71.

71. Plant 10–30 cm tall, with slightly creeping rhizome. Leaves 3–4 mm broad, nearly half as long as stem. Pistillate spikelets 1.2–1.8 (2) cm long; membranous side of sheath of bract leaves often blackish-purple. Glumes usually without light-coloured midrib............. .... 59. **C. atrofusca** Schkuhr.

+ Large plants (30) 40–80 cm tall, densely caespitose, without creeping rhizome. Leaves 5–8 mm broad, slightly shorter than stem. Pistillate spikelets 2–4 cm long; membranous side of sheath of bract leaves not purple. Glumes with distinctly bright greenish-yellow midrib .... 62. **C. griffithii** Boott.

72 (68) Plants densely caespitose, with shortened (not creeping) rhizome .... 73.

+ Plants loosely or densely caespitose, with long, less frequently short, creeping slender rhizome .... 74.

73. Leaves 3.5–6 mm broad, flat, slightly deflexed, shortly acuminate, half as long as stem or less (less often almost equal to it). Pistillate spikelets

usually drooping, lower one on up to 6 cm long peduncles, projecting significantly from sheath of bract leaf. Utricles without veins, with elongated beak..............67. **C. stenocarpa** Turcz. ex V. Krecz.

+ Leaves 2–3 mm broad flat or longitudinally folded, narrow and erect, markedly acuminate, rather sinuate at ends, slightly shorter than stem. Pistillate spikelets erect, their peduncles concealed in sheaths of bract leaves or scarcely projecting. Utricles with veins and short beak ....58. **C. alexeenkoana** Litv.

74 (72) All spikelets in inflorescence aggregated. Utricles with elongated beak..63. **C. przewalskii** Egor.

+ Upper spikelets aggregated; lower ones rather distant. Utricles usually with short beak ....75.

75. Pistillate spikelets erect, cylindrical, their peduncles not projecting (in general) from sheath of bract leaves; glumes wholly conceal utricles...................58. **C. alexeenkoana** Litv. (also see couplet 73).

+ Pistillate spikelets drooping, spreading or less often erect, ovate or elliptical, their peduncles projecting from sheaths of bract leaves by 1–2.5 (3) cm; glumes shorter than utricles.................................. ....64. **C. macrogyna** Turcz. ex Steud.

76 (67) Utricles rather densely pubescent all over surface.......................77.

+ Utricles glabrous or less often with barely perceptible pubescent beak..85.

77. Spikelets in inflorescence aggregated; pistillate spikelets ovate or spherical, on peduncles not more than 1 cm long.......................78.

+ Spikelets distant; pistillate spikelets rather oblong, on long peduncles, partly concealed in sheaths of bract leaves..............80.

78. Nut without disc on top. Staminate spikelet pale. Utricles with long beak. Leaf sheath at base of stems purplish-brown................ ....***C. amgunensis** Fr. Schmidt, p. 75.

+ Nut with disc on top. Staminate spikelet brown. Utricles with short beak. Leaf sheath at base of stems chestnut-brown.....................79.

79. Glumes acute or aristate, longer than utricles or as long............... ....42. **C. titovii** V. Krecz.

+ Glumes usually rather blunt and shorter than utricles................... ....41. **C. caryophyllea** Latourr.

80 (77) Plant with elongated slender creeping rhizome. Spikelets pale yellow. Lower bract leaf with long leaf blade.. ....50. **C. allivescens** V. Krecz.

+ Plant without creeping rhizome or with short creeping rhizome, thickened due to numerous leaf remnants. Spikelets usually brown or purplish-brown...81.

81. Plants usually loosely caespitose, with short, ramose, creeping rhizome. Shoots surrounded at base by dark purple leaf sheaths, often incised in a reticular manner. Sheath of lower bract leaf

usually without leaf blade, membranous and sometimes shortly acuminate at tip (awn 0.2–0.5 cm long) .... 82.

\+ Plant densely caespitose, without creeping rhizome. Shoot base surrounded by brown leaf sheaths. Sheath of lower bract leaf mostly with narrow or setaceous leaf blade 1–3 cm long .... 83.

82\. Staminate spikelet, as a rule, disposes above pistillate spikelets, (1) 1.5–2.5 cm long, 4–5 mm in diameter, peduncle 2–3 cm long; lower glume of staminate spikelet much less than half its length. Pistillate spikelets usually two, less often three. Utricles without or with indistinct veins .... 53. **C. macroura** Meinsh.

\+ Staminate spikelet usually at same level as upper pistillate spikelet, 0.5–1 cm long, about 1.5 mm in diameter, sessile or on 0.5–1 cm long peduncle; lower glume of staminate spikelet half as long or more, sometimes almost equal. Pistillate spikelets 3 or 4. Utricles with distinct veins .. 52. **C. lanceolata** Boott var. **alaschanica** Egor.

83 (81) Staminate spikelet (1) 1.5–2 cm long, 3–5 mm in diameter, peduncle (1.5) 2–3 cm long, much larger than pistillate spikelets .... 55. **C. supermascula** V. Krecz.

\+ Staminate spikelet 0.5–1 (1.5) cm long, on 0.5–1.5 cm long peduncle, generally at same level as upper pistillate spikelet or slightly below or above it .... 84.

84\. Utricles 3.5–4 mm long, 1.5–1.8 mm broad, with perceptibly attenuated base, with indistinct veins or none in front, with 3–5 ribbed veins at back; utricles seen well in lower half. Glumes generally light brown (Mongolia) .... 54. **C. pediformis** C. A. Mey.

\+ Utricles (2.8) 3–3.3 mm long, 1.5–1.7 mm broad, with weakly attenuated base, without veins. Glumes mostly brown (Sinkiang) .... 51. **C. aneurocarpa** V. Krecz.

85 (76) Plant with long and slender creeping rhizome. Beak of utricles and keel of glumes covered with extremely fine papillae seen well under binoculars; utricles convex-trigonous, yellowish .... 56. **C. panicea** L.

\+ Plants without creeping rhizome. Utricles and glumes without papillae .... 86.

86\. Utricles with veins. Pistillate spikelets dense, erect .... 87.

\+ Utricles without veins. Pistillate spikelets with loosely set flowers, often drooping .... 89.

87\. Spikelets in inflorescence aggregated, sessile; only lower spikelet sometimes considerably distant on long peduncle; pistillate spikelets mostly spherical-ovate. Lower bract leaf much larger than inflorescence. Glumes of pistillate spikelets acute but without awns .... 74. **C. philocrena** V. Krecz.

\+ Spikelets distant, especially in lower part; lower ones on elongated peduncles; pistillate spikelets mostly cylindrical. Lower bract leaf shorter than inflorescence or equal. Glumes of pistillate spikelets with rather scabrous awn ....88.

88. Beak of utricles smooth ....73. **C. diluta** M.B.

\+ Beak of utricles rather scabrous along margin, sometimes only with stray spines ....72. **C. aspratilis** V. Krecz.

89 (86) Staminate spikelet clavate, dense, (0.8) 1–1.5 cm long, 3–5 mm in diameter, brownish-orange. Glumes dark brown. Utricles (2.8) 3–3.5 mm long, with elongated setaceous-scabrous beak..68. **C. ledebouriana** C. A. Mey.

\+ Staminate spikelet linear, 0.5–1 cm long 2–2.5 mm in diameter, pale green or pale brown. Glumes pale or light brown or light green. Utricles orbicular-trigonous, with short, generally smooth beak..90.

90. Mature utricles greenish-brown, 2.2–2.7 (3) mm long, ovate or oblong-ovate, lustrous......69. **C. karoi** (Freyn) Freyn (also see couplet 36).

\+ Mature utricles pale green, 1.7–2 mm long, almost broadly ovate, not lustrous........71. **C. selengensis** Ivanova (also see couplet 36).

91 (66) Inflorescence capitate or capitate-spicate, consists of aggregated spikelets; staminate spikelet single; pistillate spikelets ovate or oblong-ovate. Utricles without or with indistinct veins, membranous or coriaceous. Lower bract leaf scaly or with short blade. Glumes acute or rather blunt ....92.

\+ Inflorescence not capitate; pistillate spikelets distant, cylindrical; staminate spikelets generally 2–4, less often single. Utricles, with rare exception, with distinct veins, coriaceous. Lower bract leaf with long blade reaching tip of inflorescence. Glumes aristate .... 94.

92. Plant without creeping rhizome, densely caespitose, tuft forming. Leaves about 1 mm broad, setaceous-grooved..*C. **meyerian** Kunth, p. 72.

\+ Plant with long creeping rhizomes, not forming tufts. Leaves usually flat, less often rather longitudinally folded, 3–6 mm broad....... 93.

93. Inflorescence so dense that borders between pistillate spikelets (especially in plants with immature utricles) indistinguishable and spikelets somewhat fused into a single mass. Utricles ovate or elliptical, trigonous, slightly inflated, 3–3.5 mm long, beak entire, 0.2–0.3 mm long. Leaves (3) 5–7 mm broad.......34. **C. melanantha** C. A. Mey.

\+ Inflorescence less dense, spikelets in inflorescence well distinguishable. Utricles elliptical or obovate, 2.3–4 mm long, suborbicular in section, with short and acute bicuspid beak about 0.5 mm long. Leaves 3–5 mm broad ....37. **C. moorcroftii** Falc. ex Boott.

94 (91) Utricles flattened-trigonous or nearly flat, 4 mm long, greenish-black with emarginate beak. Pistillate spikelets 2.5–7 cm long, 0.5–0.7 mm in diameter. Leaves 3–8 mm broad...........85. **C. acutiformis** Ehrh.

+ Utricles convex-trigonous, brownish-grey or orange to reddish and dark purple, with bicuspsid beak..95.

95. Large coastal and swamp plants 50–120 (150) cm tall. Leaves 4–15 mm broad, mostly knotty-reticular due to prominent transverse veins. Staminate spikelets thickened, cylindrical. Utricles ovate-conical, 5–6.5 mm long, brownish-grey, with slender prominent veins..96.

+ Plants of moderately humid habitats, mostly meadows, less tall at up to 50 cm. Leaves 2–4 mm broad, not reticular or slightly so. Staminate spikelets narrow. Utricles with sunken or prominent veins (in latter event, utricles 3–4 mm long) or without veins............97.

96. Glumes with awns up to 2.5 mm long, perceptibly longer than utricles..88. **C. riparia** Curt.

+ Glumes with very short awns or without them, acute, shorter (sometimes by half) than utricles.....................89. **C. rugulosa** Kük.

97 (95) Utricles without or with very indistinct veins, not lustrous. Spikelets in inflorescence aggregated; upper pistillate spikelet away from staminate by not more than 1 cm. Rhizome slender.... ....*asterisk **C. heterostachya** Bunge, p. 97

+ Utricles with distinct veins, lustrous or dull. Spikelets in inflorescence distant; upper pistillate spikelet away from staminate by 2 cm or more (less often, not as much). Rhizome rather thickened....... 98.

98. Utricles 5–6 mm long, brownish-grey, dull, with sunken veins. Pistillate spikelets thickened, up to 1 cm in diameter; glumes and pistillate spikelets dark brown ....87. **C. melanostachya** M.B. ex Willd.

+ Utricles 3–4 mm long, from orange to dark purple, lustrous or otherwise, with prominent veins. Pistillate spikelets 5–6 mm broad; glumes of pistillate spikelets brownish ....99.

99. Utricles generally orange, lustrous, ovate or broadly ovate, with slender prominent, but not ribbed viens.........90. **C. songorica** Kar. et Kir.

+ Utricles generally dark purple, dull or slightly lustrous, usually ovate, with thickened ribbed veins ....86. **C. gotoi** Ohwi.

Subgenus **Primocarex** Kük

Section **Petraea** (O. Lang) Kük.

1. **C. argunensis** Turcz. ex Trev. in Ledeb. Fl. Ross. 4 (1852) 267; Bess. in Flora, 17, 1, Beibl. (1834) 27 (nom. nud.); Kük. in Pflanzenreich [IV, 20], 38 (1909) 87; V. Krecz. in Fl. SSSR, 3 (1935) 382; Grubov, Consp. fl. MNR (1955) 82.

Described from East. Siberia. Type in Leningrad. Plate I, fig. 2.

On steppe slopes, sands; up to alpine belt.

IA. Mongolia: ***Mong.-Alt.*** (Tolbo-Kungei mountain range, alpine steppe, 5 VIII 1945—Yun.); ***East. Mong.*** (near Kharkhonte railway station, sand, 7 VI 1902—Litw.; Choibolsan somon, 10 km west of Khabirgi station, in Bayan-Bulak region, gentle slopes of mud volcano, herbage-fine grass steppe, 1 VI 1955—Dashnyam).

General distribution: East. Sib. (south.), North. Mong. (Hent., Mong.-Daur., Cis.-Hing.) China (Dunbei).

2. C. **obtusata** Liljebl. in Svensk. Vet. Ac. Handl. 14 (1793) 69, Table 4; Kük. in Pflanzenreich [IV, 20], 38 (1909) 87; Krylov, Fl. Zap. Sib. 3 (1929) 434; V. Krecz. in Fl. SSSR, 3 (1935) 381; Grubov, Consp. fl. MNR (1955) 86; Fl. Kazakhst. 2 (1958) 73. —**Ic.**: Russk. bot. zh. 3–6 (1911), 36, fig. 14.

Described from Sweden. Type in Stockholm.

In larch forests, alpine meadows and on steppes.

IA. Mongolia: ***Khobd. Mong. Alt.*** (nor. slopes of Khara-Tszarga mountain range near Khairkhan-Duru river, larch forest, 25 VIII 1930—Pob.; Khan-Taishiri mountain range, nor. slope, larch forest, 12 VII 1945; Indertiin-Gol—Tumurtu-Gol pass, alpine meadow, 25 VII 1947—Yun.); ***Cent. Khalkha, East. Mong.*** (Shilin-Khoto lown, chee grass pasture, 1959—Ivan.); ***Gobi-Alt.*** (Dzun-Saikhan mountain range, nor. part, upper third of nor. slope, on margin, 19 VI 1945—Yun.).

IIA. Jungar. ***Alt. Region.*** (Qinhe, Chzhunkhaitsza, 2500 m, Nos. 1131 and 1141, 5 VIII 1956—Ching); ***Jung. Alt.*** (Toli district, high-altitude meadows, 1950 m, 8 VIII 1957—Kuan).

General distribution: Balkh. Region (Kent hills), Jung.-Tarb. (Saur mountain range); Arct. (Asian), Europe [Sweden, east. Cent. Europe, Europ. USSR (stray finds), Urals], Caucasus, West. Sib. (less often, south. part and Altay), East. Sib., Far East, North Mong. (Fore Hubs., Hent., Hang.), China (Dunbei), North Amer. (west.).

3. C. **rupestris** Bell. ex All. Fl. Pedem. 2 (1785) 264; Kük. in Pflanzenreich [IV, 20], 38 (1909) 86; Krylov, Fl. Zap. Sib. 3 (1929) 433; V. Krecz. in Fl. SSSR, 3 (1935) 381; Kitag. Lin. Fl. Mansh. (1939) 110; Grubov, Consp. fl. MNR (1955) 87. —**Ic.**: All. l.c. Plate 92, fig. 1; Fl. Murm. obl. 2 (1954), Plate XLI, fig. 1.

Described from Alps. Type in ?Plate I, fig. 3.

On rock placers and talus, among rocks, in dry Kobresia meadows in high-altitude belt.

IA. Mongolia. ***Khobd.*** (on Kharkhiry peak, 24 VII 1879—Pot.), ***Mong. Alt.*** (Yangtze river, ***Gobi-Alt.*** (Ikhe-Bogdo mountain range, Narin-Khurimt ravine, middle of creek, nor. slope, among rocks, 28 VI; same site, upper belt near peak, hummocky area among rock placers, 29 VI; same site, above Ketsu creek, kobresia thicket, 25 VI; same site, flat crest of mountain range in upper courses of Bityuten river, high-altitude belt, rubble placer, 3700 m, 29 VI; same site, upper Shimkhid creek, high-altitude belt, dry Kobresia meadow, 30 VI 1945—Yun.).

General distribution: Arct., Europe (nor. and hills of West. Europe, Kola peninsula, Urals), Caucasus, West. Sib. (Altay), East. Sib. (south.), Far East, Nor. Mong. (Hent, Hang.), China (Dunbei), Nor. Amer. (west.).

Section Unciniiformes Kük.

4. C. **microglochin** Wahl. in Svensk. Vet. Ac. Handl. 24 (1803) 140; Forbes and Hemsley, Index Fl. Sin. 3 (1905) 298; Kük. in Pflanzenreich [IV, 20], 38 (1909) 108, p.p.; Krylov, Fl. Zap. Sib. 3 (1929) 435; Pampanini, Fl. Carac. (1930) 83; V. Krecz. in Fl. SSSR, 3 (1935) 302; Fl. Kirgiz. 2 (1950) 288; Grubov, Consp. fl. MNR (1955) 86; Fl. Kazakhst. 2 (1958) 66; Fl. Tadzh. 2 (1963) 118; Ikonnikov, Opred. rast. Pamira (Key to Plants of Pamir) (1963)

79. —"C. *cephalotes* Muell. var. *altior* Kük."—Danguy in Bull. Mus. nat. hist. natur. 17, 6 (1911) 451. —**Ic.**: Fl. Kazakhst. 2 (1958), Plate III, fig. 10.

Described from Sweden. Type in Stockholm. Plate I, fig. 1.

On hummocky marshes along banks of rivers and lakes and in upper hill belt.

IA. Mongolia: ***Khobd.*** (Katu river, 16 VI 1879—Pot.); ***Cent. Khalkha, Val. lakes.***

IB. Mongolia: ***West.*** (Sinkiang-Tibet highway, 27 km west of Maigaiti, No. 00481, 3 VI, 1959—Lee).

IIA. Junggar: ***Jung. Gobi, Tien Shan*** (B. Yuldus basin, 30 to 35 km south-west of Bain-Bulak settlement, basin floor, grassy swamp, along borders of a mound, 10 VIII 1958—Yun.; on B. Yuldus marsh, close to water, 2460 m, No. 6467, 10 VIII 1958—Lee and Chu).

IIIA. Qinghai: ***Nanshan*** (Shargagol'dzhin river, Baga-Bulak area, marsh, 3150 m, 15 VI 1894—Rob.; "Gachoun, pâturages du Nan-Chan, No. 675, 23 V 1908, L. Vaillant"—Danguy, l.c.).

IIIB. Tibet: ***Weitzan*** (Valley of Alyk-Norin-Gol river, Alyk-Nor lake, in hummocky marshes, along banks of lakes and rivers, 3690 m, 28 V 1900—Lad.).

General distribution: Jung.-Tarb., Nor. and Cent. Tien Shan, East. Pamir; Arct. (Europe), Europe (Scandinavia and Cent. Europe, Lit. SSR), Caucasus, Mid. Asia (West. Tien Shan—Talassk mountain range and Pamiroalay), West. Sib. (Tarsk. region and Altay), East. Sib., North Mong. (Fore Hubs., Hent. Hang., Mong.-Daur), China (North-west. and South-west.) Himalayas (west.), North Amer. (north), South Amer. (Tierra del Fuego).

Note. The plant cited by Danguy (l.c.) for Nanshan under the name "C. *cephalotes* Muell. var. *altior* Kük." belongs in all probability, to C. *microglochin*. It must be stated that there is no such combination whatsoever. There is C. *pyrenaica* var. *altior* Kük. [Pflanzenreich, 38 (1909) 106] which is a Japanese plant. C. *cephalotes*. however, grows in Australia. Reports of any one of these plants in Nanshan are clearly erroneous.

Subgenus **Vignea** (Beauv.) Peterm.

Section Vulpinae (Carey) Christ

5. **C. vulpina** L. Sp. pl. (1753) 973; Kük. in Pflanzenreich [IV, 20], 38 (1909) 198, p.p.; Krylov, Fl. Zap. Sib. 3 (1929) 446; V. Krecz. in Fl. SSSR, 3 (1935) 150; Fl. Kazakhst. 2 (1958) 49. —**Ic.**: Fl. SSSR, 3, Plate XI, fig. 13.

Described from Europe. Type in London (Linn.).

On wet and marshy sites.

IIA. Junggar: ***Altay Region*** (in Fuyun' [Koktogoi] region, 1200 m, 13 VIII 1956—Ching); ***Jung. Gobi*** (north: Ch. Irtysh, 26 VIII 1876—Pot.; 5 km south-east of Shibati, in floodplain, 580 m, No. 10443, 13 VI 1959—Lee et al.).

General distribution: Aral-Casp. (north-east.), Balkh. Region (Zaisan); Europe, Caucasus, West. Sib. (excluding Altay), East. Sib. (southern part, excluding Sayan mountain range).

Section Muhlenbergianae Tuckerm. ex Kük.

6. **C. polyphylla** Kar. et Kir. in Bull. Soc. natur. Moscou, 14 (1841) 859; V. Krecz. in Fl. SSSR, 3 (1935) 155; Fl. Kirgiz. 2 (1950) 276; Fl. Kazakhst. 2 (1958) 50; Fl. Tadzh. 2 (1963) 90. —*C. echinata* var. *leersii* (F. Schultz) Kük. in Pflanzenreich [IV, 20], 38 (1909) 161. —*C. pairaei* auct. non F. Schultz: Krylov, Fl. Zap. Sib. 3 (1929) 444, p.p. —**Ic.**: Fl. SSSR, 3, Plate XI, fig. 6.

Described from Junggar (Tarbagatai mountain range). Type in Moscow.
In shrubs and thin forests in forest belt.

IIA. Junggar: ***Tien Shan*** (Borgaty brook on northern side of Kash river, 1500–1800 m, 5 VII 1879—A. Reg.; Talki, [28] VII 1877—A. Reg.).

General distribution: Balkh. region (south-east.), Jung.-Tarb., Nor. and Cent. Tien Shan; Europe (including south. Europ. part USSR), Balk.-Asia Minor, Near East, Caucasus, Mid. Asia (Pamiroalay, West. Tien Shan, hilly Turkmenia), West. Sib. (Altay, Zmeinogorsk and Tom' river region), East. Sib. (Minusinsk region), Himalayas (Kashmir).

Section Holarrhenae (Doell) Pax

7. **C. curaica** Kunth, Enum. pl. 2 (1837) 375; Kük. in Pflanzenreich [IV, 20], 38 (1909) 124, p.p.; Krylov, Fl. Zap. Sib. 3 (1929) 440; V. Krecz. in Fl. SSSR, 3 (1935) 137; Grubov, Consp. fl. MNR (1955) 83; Fl. Kazakhst. 2 (1958) 47; Egorova in Bot. mater. Gerb. Bot. inst. AN SSSR, 19 (1959) 80. —**Ic.:** Fl. SSSR, 3, Plate X, fig. 12.

Described from Altay. Type in Berlin (?).

In marshy meadows.

IA. Mongolia: ***Khobd., Mong. Alt.*** (upper Indertiin-Gol, marshy meadow in high-altitude belt, 24 VII 1947—Yun.); ***Depr. Lakes, Val. Lakes.***

IIA. Junggar: ***Altay Region*** (in Quinhe region, 250 m, without No., 4 VIII 1956—Ching); ***Jung.-Gobi.***

General distribution: Jung.-Tarb.; West. Sib. (Altay), East. Sib. (south.), Nor. Mong. (Fore Hubs., Hent., Hang.).

8. **C. lithophila** Turcz. in Bull. Soc. natur. Moscou, 28, 1 (1855) 328; V. Krecz. in Fl. SSSR, 3 (1935) 130; Kitag. Lin. Fl. Mansh. (1939) 105; Grubov, Consp. fl. MNR (1955) 85. —*C. intermedia* auct. non Good.: Forbes and Hemsley, Index Fl. Sin. 3 (1905) 135, p.p.

Described from East. Siberia. Type in Leningrad.

In wet meadows and on marshy river banks.

IA. Mongolia: ***Mong. Alt.*** (silty-marshy lowlands in Bulugun river floodplain, 18 IX 1930—Bar.); ***Cis-Hing.*** (near Yaksha railway station, wet meadow, 13 VI 1902—Litw.); ***East. Mong.*** (near Kharkhonte railway station, interdune basin, 9 VI 1902—Litw.); ***Bas. lakes*** (Ulangom, in a marsh, 2 and 3, VII 1879—Pot.).

General distribution: East. Sib. (south.), Far East (south.), North Mong. (Hent., Hang., Mong.-Daur), China (Dunbei, North, North-west.—Gansu province).

9. **C. pycnostachya** Kar. et Kir. in Bull. Soc. natur. Moscou, 15 (1842) 522; V. Krecz. in Fl. SSSR, 3 (1935) 138; Fl. Kirgiz. 2 (1950) 272; Fl. Kazakhst. 2 (1958) 48; Fl. Tadzh. 2 (1963) 91; Ikonnikov, Opred. rast. Pamira (Key to Plants of Pamir) (1963) 79. —*C. borotalicola* Rgl. in Acta Hort. Petrop. 7 (1880) 566. —*C. curaica* var. *pycnostachya* (Kar. et Kir.) Kük. in Pflanzenreich [IV, 20], 38 (1909) 124. —*C. curaica* auct. non Kunth: Kük. l.c. 124, p.p. quoad pl. Asiae centr.; Fl. Kirgiz. 2 (1950) 272; ?Pampanini, Fl. Carac. (1930) 83. —*C. vulpinaris* auct. non Nees: Kük. l.c. 117, quoad pl. turkest. —**Ic.:** Fl. Tadzh. 2, Plate XIX, figs. 3 and 4.

Described from Junggar (Jung. Alatau). Type in Leningrad.

In marshy meadows, grass-sedge swamps, near springs; from foothills to midmontane belt.

IIA. Junggar: *Tien Shan* (Sygashu, 600 m, 4 V; from Borgata to Khapchagai, 1800–2100 m, 6 VII; midcourse of Taldy, 2100 m, 26 V 1879—A. Reg.; 20–25 km south of Urumchi, swampy short-grass meadows, 2 VI 1957—Yun. et al.; B. Yuldus basin, 30–35 km south-west. of Bain-Bulak settlement, grassy marsh near lakelets fed by springs and marshy meadows, 10 VIII 1958—Yun.; same site, Nos. 6465, 6471 and 6480, 10 VIII 1958—Lee and Chu); *Jung. Gobi* (south.: Koltun picket, 450 m, 3 V 1879—A. Reg.; 2 km nor. of Kuitun, No. 380, 6 VII 1957—Kuan; east.: Turkyul' lake, in pool of a spring, 16 VI 1877—A. Reg.).

IIIB. Tibet: *Chang Tang* ("Valle del Caracásh, 4250 m, Henderson"—Pampanini, l.c.).

General distribution: Aral-Casp. (Mugodzhary and B. and M. Barsuki), Balkh. Region (Karkaralinsk), Jung. Tarb., North and Centr. Tien Shan, East. Pam.

Section Vignea Koch

10 **C. diplasiocarpa** V. Krecz. in Fl. SSSR, 3 (1935) 590 and 135; Kitag. Lin. Fl. Mansh. (1939) 100. —*C. praecox* auct. non Schreb.: Kük. in Pflanzenreich [IV, 20], 38 (1909) 131, p.p. —**Ic.:** Fl. SSSR, 3, Plate X, fig. 9.

Described from Far East. Type in Leningrad.

In meadows.

IA. Mongolia: *Cis-Hing.* (near Yaksha railway station, wet meadow, 13 VI 1902—Litv.); *East. Mong.*

General distribution: East. Sib. (Amginsk region), Far East (south.), China (Dunbei).

***C. pallida** C.A. Mey. in Mém. Ac. Sci. St.-Pétersb. Sav. Étr. 1 (1831) 215; Kük. in Pflanzenreich [IV, 20], 38 (1909) 134, p.p.; V. Krecz. in Bot. zh. 22, 1 (1937) 109. —*C. accrescens* Ohwi in Mem. Coll. Sci. Kyoto Univ. ser. B, 6 (1931) 255; V. Crecz. in Fl. SSSR, 3 (1935) 136; Kitag. Lin. Fl. Mansh. (1939) 107; Grubov, Consp. fl. MNR (1955) 81. —*C. siccata* auct. non Dew.: Franch. Pl. David. 1 (1884) 319. —**Ic.:** C. A. Mey. l.c. Tab. VIII; Fl. SSSR, 3, Plate X, fig. 11.

Described from East. Siberia. Type in Leningrad.

In forest glades and under forest canopy.

Found at boundary of Cis-Hing., Mongolia region (near Irekte railway station, forest glade, 17 VI; near Hinggan railway station, marshy forest, 19 VI 1902, Litw.).

General distribution: Arct. (Asiat.), East. Sib., Far East, North Mong. (Hent.), China (Dunbei), Korea (north).

Note. The reference by Walker [Contribs. U.S. Nat. Herb. 28 (1941) 599] to the presence of *C. pallida* in Alashan Gobi (in dry sandy soil and on slopes) is erroneous and, in all probability pertains to *C. stenophylloides*, which is common there.

11. **C. praecox** Schreb. Spicil. Fl. Lips. (1771) 63; Kük. in Pflanzenreich [IV, 20], 38 (1909) 129; Krylov, Fl. Zap. Sib. 3 (1929) 441; V. Krecz. in Fl. SSSR, 3 (1935) 135; Fl. Kazakhst. 2 (1958) 47. —**Ic.:** Fl. SSSR, 3, Plate X, figs. 8 and 8a.

Described from Europe. Type in München.

In meadow steppes.

IIA. Junggar: ***Alt. Region*** (Shara-Sume, on ravine floor, No. 2623, 29 VIII 1956—Ching; 15–20 km nor.-west of Shara-Sume, scrub steppe, 7 VII; 20 km west of Shara-Sume on Kran river, scrub meadow steppe, 7 VII 1959—Yun.).

General distribution: Aral-Casp., Balkh. Region (east.), Jung.-Tarb.; Europe, Caucasus, West. Sib., East. Sib.

Section Boernera V. Krecz. ex Egor.

12. **C. duriuscula** C. A. Mey. in Mém. Ac. St.-Pétersb. Sav. Étr. 1 (1831) 214; V. Krecz. in Fl. SSSR, 3 (1935) 140; Kitag. Lin. Fl. Mansh. (1939) 100; Grubov, Consp. fl. MNR (1955) 83; Egorova in Bot. mater. Gerb. Bot. inst. AN SSSR, 19 (1959) 56. —*C. stenophylla* var. *duriuscula* Trautv. in Acta Horti. Petrop. 10 (1887) 537; Kük. in Pflanzenreich [IV, 20], 38 (1909) 121; Krylov, Fl. Zap. Sib. 3 (1929) 439. —*C. stenophylla* auct. non Wahl.: Franch. Pl. David. 1 (1884) 319; Forbes and Hemsley, Index Fl. Sin. 3 (1905) 312; Danguy in Bull. Mus. nat. hist. natur. 20 (1914) 145; Egorova, l.c. 50, p. min. p., quoad pl. mongol. —*C. uralensis* auct. non Clarke: Grubov, l.c. 89. —**Ic.:** C.A. Mey. l.c. Tab. VIII; Fl. SSSR, 3, Plate X, fig. 3.

Described from East. Siberia. Type in Leningrad.

On sandy sites: in steppes that sometimes have become deserts, along steppe slopes, in meadows, not infrequently saline, along banks of rivers and brooks, along rock and rubble slopes; on plains and up to mid-montane belt.

IA. Mongolia: ***Khobd.*** (Altyn-Khatysyn area, along river bank, 18 VI; valley of Kharkhiry river, on sandy soil, 21 VII 1879—Pot.; valley of Sagli river, between sand-dunes, 4 X 1931—bar.); ***Mong. Alt., Cis-Hing.*** (near Yaksha railway station, 13 VI 1902—Litw.); ***Cent. Khalkha, East. Mong., Depr. lakes*** (Ulangom, in meadow along brook, 3 VII; Kholbo-nur lake, on sandy soil, 27 VII 1879—Pot.); ***Val. Lakes*** (east. border of extensive Tuilin-tala plain, chee grass steppe, on sandy loams, 26 VIII; 5 km east of Bain-Khongor camp on western road between Tsagan-Ol and Tszag-Baidarik, chee grass steppe, 27 VIII 1943—Yun.; 20 km nor.-nor.-west of Delger, chee grass steppe, 30 VIII 1948—Yun.); ***Gobi-Alt.*** (Gurbun Saikhan, rocky canyon banks, 1950 m, 1925—Chaney; Ikhe-Bogdo mountain range, wormwood-chee grass hill steppe, 10 IX 1943—Yun.; Bain-Tsagan somon, east. extremity of Gichigine-nut mountain range, onion-chee grass steppe, 27 VIII 1948—Grub.); ***East. Gobi*** (Khara-Airik somon, 10 km nor.-east of somon, saline steppe desert, 26 VIII 1940—Yun.; Delger-Khangai mountain range, chee grass desert steppe, 10 V 1941; south. spurs of Ikhe-Narata, desert steppe, 31 V 1941—Yun.; Kholt somon, plain 6 km nor.-east of Oldakhu-khid monastery, chee grass steppe, 1250 m, 15 VIII 1950—Kal.).

General distribution: Arct. (Chukchi), West. Sib. (south.); East. Sib. (predominantly south.), Far East (south.), North Mong., China (Dunbei, North), Korea (north), North Amer.

13. **C. physodes** M.B. in Mém. Soc. natur. Moscou, 2 (1809) 104; Kük. in Pflanzenreich [IV, 20], 38 (1909) 119; V. Krecz. in Fl. SSSR, 3 (1935) 194; Fl. Kazakhst. 2 (1958) 54; Fl. Tadzh. 2 (1963) 97. —**Ic.:** Fl. SSSR, 3, Plate XIII, fig. 7.

Described from Transvolga. Type in Leningrad.

On hummocky semifixed sands, in steppe deserts.

IIA. Junggar: ***Tien Shan*** (Togustarau, IV 1879—A. Reg.); ***Jung. Gobi*** (nor.: near Ulyungur lake, Shazga, 540 m, No. 10296, 12 V 1959—Lee et al.); ***Zaisan*** (left bank of Ch. Irtysh river,

Dzhelkaidar area, clayey slopes, 8 VI 1914—Schischk.; same site, 2 km south-west of Burchum settlement on road to Zimunai, steppe desert, 10 VII 1959—Yun.); *Dzhark.* (Ili river west of Kul'dzha, V 1877—A. Reg.; between Suidun and Ili, 4 V; Talki river near Suidun, 4 V; Suidun, 8 V 1878—A. Reg.; valley of Ili river 7–8 km south-west of Suidun, hummocky semifixed sands, 31 VIII 1957—Yun.).

General distribution: Aral-Casp., Balkh. Region; Europe (south-east. Europ. USSR—Ryn-Peski and sands along Kuma river), Near East, Mid. Asia (west.—sandy regions).

***C. pachystylis** J. Gay in Ann. Sci. natur. Paris, sér. 2, 10 (1838) 301; V. Krecz. in Fl. SSSR, 3 (1935) 197; Fl. Kirgiz. 2 (1950) 277; Fl. Kazakhst. 2 (1958) 54; Fl. Tadzh. 2 (1963) 98. —*C. stenophylla* var. *desertorum* Litv. in Allg. Bot. Zeit. 5, 1 (1899) 56; Kük. in Pflanzenreich [IV, 20], 38 (1909) 122. —*C. stenophylla* f. *pachystylis* (J. Gay) Kük. l.c. 121. —**Ic.:** Fl. SSSR, 3, Plate XIII, fig. 8.

Described from Near East. Type in Paris.

Along dry melkozem (silt) slopes in foothills and on submontane flats and in clayey semidesert.

Cited by V.I. Kreczetowicz (l.c.) for Kul'dzha (Jung.: Dzhark.). Although herbarium material that can confirm this report is lacking in the Herbarium of the Botanical Institute, the occurrence of this species in the western part of Chinese Junggar is possible.

General distribution: Balkh. Region, North and Centr. Tien Shan (extremely rare), Asia Minor and Near East, Caucasus, Mid. Asia (montane), India (Khojak Pan), Afr. (North).

14. **C. stenophylloides** V. Krecz. in Fl. Turkm. 1, 2 (1932) 230; idem, Fl. SSSR, 3 (1935) 141; Fl. Kirgiz. 2 (1950) 272; Egorova in Bot. mater. Gerb. Bot. Inst. AN SSSR, 19 (1959) 52; Fl. Tadzh. 2 (1963) 94; Ikonnikov, Opred. rast. Pamira (Key to Plants of Pamira) (1963) 81 —*C. dimorphotheca* Stschegl. in Bull. Soc. natur. Moscou, 27 (1854) 206 (pl. monstr.); Fl. Kazakhst. 2 (1958) 48. —*C. stenophylla* var. *rigescens* Franch. Pl. David. 1 (1884) 318. —*C. stenophylla* var. *pellucida* and var. *interrupta* Litv. in Tr. Bot. muz. 7 (1910) 84, 85, —*C. duriusculiformis* V. Krecz. in Fl. SSSR, 3 (1935) 591; Fl. Kirgiz. 2 (1950) 275; Fl. Kazakhst. 2 (1958) 49; Persson in Bot. Notis. (1938) 276. —*C. rigescens* (Franch.) V. Krecz. in Fl. SSSR, 3 (1935) 142 and 592; Kitag. Lin. Fl. Mansh. (1939) 109; Egorova in Bot. mater. Gerb. Bot. inst. AN SSSR, 19 (1959) 55. —*C. discessa* V. Krecz. ex Grub. in Bot. mater. Gerb. Bot. inst. AN SSSR, 17 (1955) 7; Grubov, Consp. fl. MNR (1955) 83.—*C. stenophylla* auct. non Wahl.: Henderson and Hume, Lahore to Yarkand (1873) 338; Hemsley in J. Linn. Soc. (London) Bot. 30 (1894) 119; ej. Fl. Tibet (1902) 202; Kük. in Pflanzenreich [IV, 20], 38 (1909) 120, p.p.; Ostenfeld in Hedin, S. Tibet, 6, 3 (1922) 91; Pampanini, Fl. Carac. (1930) 83; Walker in Contribs. U.S. Nat. Herb. 28 (1941) 600. —*C. pallida* auct. non C. A. Mey.: ?Walker, l.c. 500 —**Ic.:** Fl. SSSR, 3, Plate X, fig. 2; Fl. Tadzh. 2, Plate XX, fig. 3.

Described from Junggar (Jung. Alatau). Type in Leningrad. Map 5.

On sandy sites: steppe-covered sandy deserts, in hill steppes, on permanent sands, on saline meadows, in depressions, short-grass meadows along irrigation ditches, along river valleys, often on dried river-beds on alluvial formations (loesses); from submontane flats to high-altitude regions.

IA. Mongolia: ***Mong. Alt., Cent. Khalkha*** (48 km north-east of Sangin-Dalai on old Dalan-Dzadagad to Ulan-Bator road, 16 V 1941—Yun.); ***Depr. Lakes, Val. Lakes*** (Nurin-Khoitu-tala lowland on Bain-Tsagan somon to Delger somon road, chee grass-scrub desert steppe, 28 VIII 1948—Grub.); ***Gobi-Alt., East. Gobi, Alash. Gobi, Khesi*** (Dzakha-Dolon, 13 VII; Loukhushan' hills, along sand fringes, 17 VII 1908, Czet.; 50 km nor. of Lanzhou town, wormwood-grass semidesert, 29 VI 1957—Petr.; Uvei town, south. border of Tengeri sands, on compacted sites between knolls, 23 VI; 40 km east of Chzhan'e town, foothill plain with dry river-beds at Nanshan, 10 VIII 1958—Petr.).

IB. Kashgar: ***North*** (Uchturfan, along side of irrigation ditches, 17 V; near Uchturfan, Kukurtuk gorge, 29 V 1908—Divn.); ***West., South.*** (between Keriya and Niya, in Koshkulak, 3000 m, No. 00261, 10 VI 1959—Lee et al.); ***East.*** (Lyaodun', in a wet site, No. 465, 21 V 1957—Kuan; nor.-west of Toksun, near spring, No. 7365, 20 VI; 10 km north of Karashar, near water, 1200 m, No. 6876, 29 VII 1958—Lee and Chu).

IIA. Junggar: ***Altay Region*** (Koktogoi, 1470 m, No. 10389, 6 VI 1959—Lee et al.; 2–3 km south-east of Shara-Sume, steppe on rocky slope, 7 VII 1959—Yun.) ***Tarb.*** (50–52 km nor. of Kosh-Tologoi settlement of Khobuk river, east. trail of Saur mountain range, dry slopes, 4 VII; east. trails of Saur mountain range on Karamai-Altay road, hill steppe, 4 VII 1959—Yun.); ***Jung. Alt*** (Dzhair mountain range, 25–27 km nor.-east of Toli settlement, wormwood desert steppe, 5 VIII 1957—Yun.); ***Tien Shan, Jung. Gobi*** (except cent.), ***Zaisan*** (left bank of Ch. Irtysh 42–45 km above Burchun settlement, winterfat thicket along pebble ridges, 9 VII; 19–20 km south-west of Burchum settlement, dry pea shrub-wheat-grass steppe, 10 VII 1959—Yun.); ***Dzhark.*** (Khorgos river in Kul'dzha region, 22 IV 1877; Talki river near Suidun, 4 V 1878—A. Reg.).

IIIA. Qinghai: ***Nanshan*** (San'chuan' locality, on valley floor, loess with sand, 7 VI; valley of Lanchzha-lunva river, on dry rocky soil, 14 V 1885; Tashitu river, 25 V 1886—Pot.; Humboldt mountain range, Kuku-usu river, wet meadow, 2700–3000 m, 12 V 1894—Rob.).

IIIB. Tibet: ***Chang Tang*** ("Head of Lower Karakash Valley, 4200–4500 m, 1870" Henderson and Hume, l.c.; "Tibet, close to water at 4860 m, Thorold"—Hemsley, l.c. 1894; "Inner or eastern Tibet, near Camp XLIV [33°32′ N lat., 88°52′ E long.], 5127 m, 15 VIII 1901, Hedin"—Ostenfeld, l.c.); ***South.*** ("SW Tibet, on the road from Camp CCIII [Dara-Sumkor, 4831 m] [30°16′ N lat., 82°30′ E long.] to Camp CCIV [Bak-gyäyorap, 4870 m] [30°24′ N lat., 82°27′ E long.], 16 VII 1907, S. Hedlin"—Ostenfeld, l.c.).

IIIC. Pamir (Chumbus river, Kosh-Terek area, 7 VI 1909—Divn.; upper Kaplyk river, 4500–5000 m, 14 VII; upper Lapet river, 20 VII 1942 Serp.; 1–2 km west of Tashkurgan town, 3150 m, No. 00303, 13 VI; 3–4 km south of Tashkurgan town, rocky slopes, 3200 m, No. 00337, 13 VI 1959—Lee et al.; same site, Tashkurgan valley, on short-grass meadows near irrigation ditches, 13 VI 1959—Yun.; "Tashkorghan Dafdar, ca. 3510 m, 30 VI 1935"—Persson, l.c.).

General distribution: Aral-Casp., Balkh. Region (south.), Jung.-Tarb., North and Centr. Tien Shan, East. Pamir; Asia Minor, Near East, Caucasus, Mid. Asia, China (North, North-east., North-west.), Himalayas (west., Kashmir).

Note. Plants intermediate in characteristics between *C. stenophylloides* and *C. duriuscula* are found in Mong. and Alt. Gobi, Bas. lakes and East. Gobi.

Our earlier doubts about the independent species status (Egorova, l.c. 1959) of plants placed by V.I. Kreczetowicz among *C. rigescens* (Franch.) V. Krecz. have been confirmed in that *C. rigescens* is identical to *C. stenophylloides* (see Egorova, Osoki SSSR [Sedges of the USSR] 1966, p. 129).

Section Foetidae Tuckerm. ex Kük.

15. **C. enervis** C. A. Mey. in Ledeb. Fl. Alt. 4 (1833) 209; V. Krecz. in Fl. SSSR, 3 (1935) 84; 192; Kitag Lin. Fl. Mansh. (1939) 101; Grubov, Consp. Fl. MNR (1955) 84; Fl. Kazakhst. 2 (1958) 55; Fl. Tadzh. 2 (1963) 102. —*C. stenophylla* var. *enervis* (C.A. Mey.) Kük. in Pflanzenreich [IV, 20], 38 (1909) 122; Krylov, Fl. Zap. Sib. 3 (1929) 439. —*C. vulpinaris* f. *angustifolia* Kük. l.c. 117, p.p., quoad pl. sibir.; Krylov, l.c. 437. —*C. similigena* V. Krecz. in Fl. SSSR, 3 (1935) 596, 193; Fl. Kirgiz. 2 (1950) 277. –**Ic.:** Ledeb. Fl. Alt. 4, Tab, 349; Fl. SSSR, 3, Plate XIII, figs. 4 and 6.

Described from Altay. Type in Leningrad.

In wet and marshy, sometimes saline meadows, grass-sedge swamps, along banks of rivers and brooks; predominantly in midmontane belt.

IA. Mongolia: ***Khodb., Mong. Alt.*** (Tsitsirin-Gol. on wet sand, 10 VII 1877—Pot.; Bodunchi river, hummocky bank, 19 VII 1898—Klem; Khasagtu-Khairkhan hills, Dundu-Tserengol river bank, on pebble bed, 17 IX 1930—Pob.; Khan-Taishiri nor. slope near crest, 21 IX 1945—Leont'ev; 2–3 km south-east of Yusun-Bulak, solonchak lowland, among woody creepers of derris, 13 VII 1947—Tuvanzhab; upper Indertiin-Gol, marshy meadow in high-altitude belt, 24 VII 1947—Yun.; east. bank of lake Tonkhil'-Nur, meadow solonchaks, 16 VII 1947—Yun.; north-east. bank of lake Tonkhil'-Nur, solonchak meadow, 7 IX 1948—Grub.); ***Cis.-Hing.*** (5 km west of Togë-Gol river, sedge meadow, 7 VIII 1949—Yun. Khalkha-Gol somon, Salkhit hills region, floodplain of Dege-Gol river, sedge marshy meadow, 10 VI 1956—Dashnyam); ***Cent. Khalkha*** (Gal-Shara somon, Gashyunyi-Gol area, sand bed, 20 V 1944—Yun.; inter-stream area between Uber- and Ara-Dzhirgalante rivers, hummocky sands with willow thickets, 2 VII 1949—Yun.); ***East. Mong.*** (10 km west of Yugodzyr, rather saline meadow, 14 V 1944—Yun.; ***Depr. Lakes*** (Ulangom, in swamp and in meadow along river, 2–3 VII 1879—Pot.); ***Val. Lakes*** (Bain-Gobi somon, Tsagan-Gol river near somon camp, grass plot with young sedge on river bank, 27 VII 1948—Grub.); ***Gobi-Alt.*** (Dundu-Saikhan hills, bank of brook, 2 VII 1909—Czet.; Dzun-Saikhan hills, on wet meadow at source of brook, 23 VIII 1931—Ik.-Gal.; pass between Dundu- and Dzun-Saikhan, along ravine floor, 22 VII 1943—Yun.; 1 km east of Dalan-Dzadagad town, rather saline meadow along brook, V–VII 1939—Surmazhab; Khongor somon, Tergetu-khuduk area, rather saline meadow, 20 VI; Bain-Tsagan mountain range, 10 km nor.-west of Bain-Bulak, rather saline meadow, 7 VII 1945—Yun.); ***East. Gobi*** (Shabarakh-Usu [Bain-Dzak], in marshes, at 1110 m, No. 87—Chaney); ***West. Gobi*** (Tsagan-Bogdo mountain range, Tsagan-Bulak area, saline meadow near spring, 1 VIII; same site, Suhzhi-Bulak spring, rather saline meadow, 4 VIII 1943—Yun.).

IIA. Junggar: ***Alt. Region*** (Qinhe district, Chzhuikhaitsza, dry meadow, 2460 m, 6 VIII 1956—Ching); ***Tien Shan*** (Nanshankou, near springs and swamps, 26 V 1877—Pot,; Davachin trough, Ulumbai area, 20–25 km south of Urumchi, marshy short-grass meadows, 2 VI 1957—Yun.).

General distribution: Balkh. Region, Jung.-Tarb. (Tarbagatai), North and Centr. Tien Shan, Mid. Asia (Pamiroalay and West. Tien Shan), West. Sib. (Altay and north spur of Kuznets Alatau), East. Sib. (including Sayans where, however, it is rare), North Mong., China (Dunbei).

Note. Utricles of plants from Gobi Altay and Western Gobi bear some distinct veins. With respect to all other characteristics, however, these specimens are indistinguishable from *C. enervis* (see also note to *C. roborowskii*).

16. **C. pseudofoetida** Kük. in Bot. Tidsskr. 28 (1908) 225; idem in Pflanzenreich [IV, 20], 38 (1909) 115; Ostenfeld in Hedin, S. Tibet, 4, 3 (1922) 91; V. Krecz. in Fl. SSSR, 3 (1935) 191; Fl. Kirgiz. 2 (1950) 277; Grubov, Consp. fl. MNR (1955) 87; Fl. Tadzh. 2 (1963) 100; Ikonnikov, Opred. rast.

Pamira (Key to Plants of Pamir) (1963) 79. —*C. slobodovii* V. Krecz. in Fl. SSSR, 3 (1935) 595, 191. —*C. incurva* auct. non Lightf.: Deasy, in Tibet and Chin. Turk. (1901) 399; Hemsley, Fl. Tibet (1902) 21; Forbes and Hemsley, Index Fl. Sin. 3 (1905) 291; Pampanini, Fl. Carac. (1930) 83. —**Ic.:** Kük. l.c. 1908, fig. 1.

Described from Tien Shan (Tersk Alatau). Type in Leningrad. Map 6.

In wet and marshy, quite often saline meadows, along banks of rivers and brooks, in silt sections among rubble and stone placers in high-altitude belt; gregarious at some places.

IA. Mongolia: ***Mong. Alt.*** (Tolbo-Kungei mountain range, kobresia meadow, 3200 m, 5 VIII 1945—Yun.; Adzhi-Bogdo mountain range, ascent in Burgastyin pass, hill steppe, 6 VIII; same site, Ara-Tszuslan—Ikhe-Gol interstream region, high-altitude belt, rubble-stone placers, 7 VIII 1947—Yun.); ***Gobi-Alt.*** (Ikhe-Bogdo mountain range, south. slope, upper Narin-Khurimt creek, stone placer, high-altitude belt, 6 IX 1943—Yun.; same site, Narin-Khurimt-ama and Ketsu-ama interstream region, 28 and 29 VI 1945—Yun.; same site, upper Narin-Khurimt gorge, 3500 m, 29 VII 1948—Grub.).

IB. Kashgar: ***Nor.*** (Kucha, intermontane basin, 2590 m, No. 10061, 26 VII 1959—Lee et al.); ***West.*** (Turugart settlement at border with USSR, on river bank, 3500 m, 20 VI 1959—Yun.; on Kashgar-Torugart highway, along bank of irrigation ditch, 3500 m, No. 09744, 20 VI 1959—Lee et al.).

IC. Qaidam: ***Mount.*** (Rittera mountain range, Baga-Khaltyn-Gol river, 3900–4200 m, 26 VI 1894—Rob.).

IIA. Junggar: ***Tien Shan*** (Muzart pass, 3000–3450 m, VIII 1877—A. Reg.; Kumbel', 2700–3000 m, 31 V; Mongili [Kash river], 3 VII 1879—A. Reg.; 7–8 km south of Danu, 3450 m, No. 524, 22 VII 1957—Kuan; Manas river basin, upper Danu-Gol river, 2–3 km above diversion to Se-daban pass, marshy sedge meadow, 22 VII 1957—Yun.; B. Yuldus high-altitude basin, 30–35 km south-west of Bain-Bulak settlement, grassy-sedge marsh, 10 VIII 1958—Yun.).

IIIB. Tibet: ***Chang Tang*** (Keriisk hills, Kyuk-Egil' river, on rocky sites, 3750–3900 m, 11 VII 1885—Przew.; Muzlyk-atasy mountain range, on sandy-solonetz, 3300–3900 m, 15 VIII; Przewalsky mountain range, north slope, on wet sites, 4200 m, 19 VIII 1890—Rob.; left bank of Karakash river, upper Khotan-Darya river, 15 km east of Kirgiz-Dzhangil pass on road to Shakhidulla, saline meadow, 2 VI; Raskem-Darya river valley near Muzar settlement, saline meadow, 4 VI 1959—Yun.; east of Karakash pass, gorge, 4500 m, No. 00469, 2 VI 1959—Lee et al.; "34°55′ N lat., 82°15′ E long., 4890 m, 19 VII 1896"—Deasy, l.c.; "Valle Shaksgam, 4825 m, Clifford"—Pampanini, l.c.).

IIIC. Pamir.

General distribution: North and Centr. Tien Shan, East. Pamir; Near East, Mid. Asia (Pamiroalay, West. Tien Shan), West. Sib. (Altay), Himalayas (west.), China (South-West.—Sichuan).

17. **C. reptabunda** (Trautv.) V. Krecz. in Izv. Bot. sada AN SSSR, 30 (1932) 134; idem, Fl. SSSR, 3 (1935) 190; Kitag. Lin. Fl. Mansh. (1939) 109; Grubov, Consp. fl. MNR (1955) 87. —*C. stenophylla* var. *reptabunda* Trautv. in Acta Hort. Petrop. 1, 2 (1872) 194. —**Ic.:** Fl. SSSR, 3, Plate XIII, fig. 3.

Described from Nor. Mongolia. Type in Leningrad.

In wet and marshy solonchak and solonetz-like meadows, solonchak swamps, banks of saline lakes.

IA. Mongolia: ***Cis-Hing.*** (near Yaksha railway station, saline marsh, 19 VIII Litw.; same site, on marsh, No. 2194, 1954—Wang); ***Cent. Khalkha*** (5 km south-east of Choiren, solonchak derris thicket, 22 VIII 1940; 45–50 km south-south-west of Sorgol-Khairkhan on old road from Ulan-Bator to Dalan-Dzadagad, solonchak-like derris thicket, 16 VII 1943; Ulan-Bator—Tsetserlik road, sand along south. fringe of Tsagan-Nur lake, solonchak-like grass plot, 25 VI 1948; 5 km east of upper bridge on Kerulen on Ulan-Bator—Undurkhan road, 22 VII 1949—Yun.); ***East. Mong.*** (Gurbunei-Bulak, 1870—Lomonosov, type!; near Khorkhonte railway station, wet meadow, 15 VI 1903—Litw.; same site, solonchak near soda lake, 10 IX 1927—Gordeev; Dariganga, Ikhe-Bulak district, Uizangin-Gobi, 23 VIII 1927—Zam.; near Manchuria railway station, solonetz-like wet site, No. 367, 25 VI; near Sinbaerkhuyunchi district, wet site near river, 29 VI 1951—S.H. Li et al.; floodplain of Kerulen river 8 km south of Engershand, 5 VI; Khamar-Daban region, 19 VI; Tamtsaga region, solonchak lowland, 23 VI 1954, Dashnyam; Khalkha-Gol somon, floodplain of Mogoityn-Gol river, herbage-sedge meadow, 12 VI 1956—Dashnyam); ***Val. lakes*** (Mogoit-Khuduk, on sand beds of monsoon streams, 17 VIII 1925—Glag.); ***East. Gobi*** (45 km nor.-east of Khan-Bogdo, solonchak-like tender sedge meadow, 22 IX 1940; 10–12 km south of Mandokhu somon, solonchak-like grass plot, 6 VI 1941; Udur-Shili somon, Toli-Bulak area, solonchak-like meadow along bank of brook, 27 VII 1946—Yun.), ***Alash. Gobi, Ordos*** (25 km south-east of Otok town, solonchak meadow near Khaolaitunao lake, 1 VIII; 20 km east of Otok town, meadow in lake basin of Ulantsaidenmao lake, 2 VIII; 50 km from Khangin town, solonchak meadow near Yan'khaitszy lake, 6 VIII; 20 km west and 50 km south of Dzhasak town, in meadows, 16 and 17 VIII 1957—Petr.).

General distribution: East. Sib. (south.), North Mong. (Fore Hubs., Hang., Mong.-Daur.), China (Dunbei, North.).

18. **C. roborowskii** V. Krecz. in Bot. mater. Gerb. Bot. inst. AN SSSR, 7, 2 (1937) 36 —C. *arenicola* auct. non Fr. Schmidt: ?Danguy in Bull. Mus. nat. hist. natur. 17, 6 (1911) 451.

Described from Tibet (Nanshan). Type in Leningrad.

Along wet and marshy solonetz-like meadows, swamps, banks of brooks and lakes; up to 3600 m above sea level.

IA. Mongolia: ***Alash. Gobi*** (in central bend of Huang He river [near Mandaltyn-Bulak well], 23 V 1872—Przew.; Dyn'-yuan'-in oasis [Bayan-Khoto], along banks of brooks, 6 IX 1908—Czet.); ***Khesi*** (between Gaotai and Fuintin, along irrigation canals and in swamp, 6, 12 and 17 VI 1886—Pot.; Sachzhou oasis, wet meadow, 1010 m, 9 IV 1894—Rob.).

IB. Qaidam: ***Plains*** (south.: khyrma zone of Barun-tszasak, on sand near water and in solonchak marshes, 2580 m, 6 V 1900—Lad.; north.-west.: Altyntag mountain range, 25 km south of Dantszinsankhou settlement, bank of Khuakhaitsza lake, 2 VIII 1958—Petr.); ***Hilly*** (Kurlyk area, along wet solonetz-like meadow, 8 and 15 V; Sondzhin-Gol spring brook, along swamp, 3000 m, 7 VI; Ichegyn-Gol river, on bank, 3000 m, 23 VI 1895—Rob.).

IIIA. Qinghai: ***Nanshan*** (Sharagol'dzhin river, Buklu-Tologoi area, sandy wet meadow, 3000 m, 14 VI, type; same site, in swamp, 14 VI; same site, Baga-Bulak area, on wet sand, 3000 m, 15 VII 1894—Rob.; "Gachoun, pâturages du Nan-Chan, 22 V 1908, Vaillant"—Danguy, l.c.); ***Amdo*** (upper course of Huang He river, 12 IV 1880—Przew.).

IIIB. Tibet: ***Weitzan*** (Burkhan-Budda mountain range, 3000–3600 m, 20 V 1884—Przew.).

General distribution: China (?North, North-west. and ?South-west.).

Note: C. *roborowskii* is possibly morphologically a very poorly distinguished geographic race. It differs from C. *enervis* V. Krecz. (l.c.) in greater stiffness of vegetative parts, broader leaves, longer glumes with very broad, white membranous margin and coriaceous (parchment-like) utricles with short beak and veins on both sides. A study of type specimens of C. *roborowskii* showed that even these do not exhibit all the aforesaid characteristics. For example, the texture of the utricles and length of beak as well as the breadth of leaves are the same in these specimens as in *C. enervis*. In most of the specimens relegated to *C. roborowskii* by V. Kreczetowicz, the distinctive characteristics of the species are manifest to an even lesser extent. They also include specimens (for example, from Qaidam and Qinghai—Baga-Bulak area) which are altogether indistinguishable from type specimens of *C. enervis*.

Notwithstanding the above situation, we have refrained from identifying *C. roborowskii* with *C. enervis* since the herbarium specimens of the former (more correctly, plants from the aforesaid regions of Central Asia) are still quantitatively inadequate to express a definite opinion about their independent status as a species. As several of the specimens do not possess mature utricles, it is not known to what extent the presence of veins in them is a characteristic feature. If this characteristic is indeed specific for *C. roborowskii*, the latter should include, evidently, the plants from Gobi Altay and Western Gobi, which have been designated as *C. enervis* in this volume.

Plants erroneously identified by Danguy (l.c.) as *C. arenicola* Fr. Schmidt—a species found on the Pacific coast—belong it would seem to *C. roborowskii* or, possibly, even *C. stenophylloides*.

19. **C. sajanensis** V. Krecz. in Izv. Bot. sada AN SSSR, 30 (1932) 133; idem, Fl. SSSR, 3 (1935) 190; Grubov, Consp. fl. MNR (1955) 88; Hanelt and Davažamc in Feddes Repert, 70, 1–3 (1965) 17. —**Ic.:** Fl. SSSR, 3, Plate XIII, fig. 2.

Described from East. Siberia (East. Sayan mountain range). Type in Leningrad.

On sands and pebble beds along banks of water reservoirs.

IA. Mongolia: ***Khobd.*** (Katu river, on dry pebble bed, at valley bottom, 16 VI 1879—Pot.); ***Mong. Alt.*** (Tsitsirin-Gol, on wet sand, near water, 18 VII 1877—Pot.); ***Cent. Khalkha*** (Kholt area in Hangay foothills, 28 V 1926—Gus.), ***East. Mong., Val. lakes*** ("NW Ufer des Orognur, Solontschakwiese, No. 2146, 1962"—Hanelt and Davažamc, l.c.); ***East. Gobi*** (Lus somon, Khatu-Tugrik mountain range, hummocky trough with lake remnants, on sand, 17 VI 1950—Kal.).

General distribution: East. Sib., North Mong. (Fore Hubs., Hent., Mong.-Daur.).

Section Canescentes (Fries) Christ

20. **C. canescens** L. Sp. pl. (1753) 974; Kük. in Pflanzenreich [IV, 20], 38 (1909) 216, p.p.; Krylov, Fl. Zap. Sib. 3 (1929) 452; V. Krecz. in Fl. SSSR, 3 (1935) 176; Kitag. Lin. Fl. Mansh. (1939) 98; Fl. Kirgiz. 2 (1950) 276; Grubov, Consp. fl. MNR (1955) 82; Fl. Kazakhst. 2 (1958) 53 —**Ic.:** Fl. SSSR, 3, Plate XII, figs. 8 and 9.

Described from North Europe. Type in London (Linn.).

In marshy meadows.

IA. Mongolia: ***Mong. Alt.***

IIA. Junggar. ***Altay Region*** (Qinhe [Chingil'] river, Chzhunkhaitsza, Nos. 1389, 1404 and 1410, 6 VIII 1956—Ching); ***Tien Shan*** (Mengute, 2700 m, 2 VIII 1879—A. Reg.).

General distribution: Jung.-Tarb., Nor. and Cent. Tien Shan; Arct., Europe, Caucasus, Mid. Asia (West. Tien Shan and Angren river), West. Sib., East. Sib. (excluding Sayan mountain range), Far East, North Mong. (Hent., Mong.-Daur.), China (Dunbei), Himalayas (Kashmir), Korea, Japan, North America, South America, Austral.

21. **C. tripartita** All. Fl. Pedem. 2 (1785) 265; V. Krecz. in Fl. SSSR, 3 (1935) 181; Kitag. Lin. Fl. Mansh. (1939) 112. —*C. lachenalii* Schkuhr, Riedgr. 1 (1801) 51; Krylov, Fl. Zap. Sib. 3 (1929) 450. —*C. lagopina* Wahl. in Svensk. Vet. Ac. Handl. 24 (1803) 145; Kük. in Pflanzenreich [IV, 20], 38 (1909) 213. —**Ic.**: Fl. SSSR, 3, Plate XII, fig. 13.

Described from Europe (Alps). Type in Berlin?

In high-altitude zone.

IIA. Junggar: ***Altay Region*** (Altay hills in Timulbakhana region, in high-altitude zone, 2600 m, No. 10704, 19 VII 1959—Lee et al.).

General distribution: Arct., Europe (hilly regions), West. Sib. (basin of Nor. Sos'va and Altay), East. Sib., Far East, China (Dunbei, Altay), Korea, Japan, India (Assam—Jaintia hills), Nor. Amer., New Zealand.

Subgenus **Carex**

Section Acutae (Fries) Christ

22. **C. appendiculata** (Trautv. et C. A. Mey.) Kük. in Bull. Hérb. Boiss. 2 sér. 4 (1904) 54; idem, Pflanzenreich [IV, 20], 38 (1909) 338; V. Krecz. in Fl. SSSR, 3 (1935) 214; Grubov, Consp. fl. MNR (1955) 82. —*C. acuta* var. *appendiculata* Trautv. et C. A. Mey. in Middend. Sibir. Reisen. 1, 2 (1956) 100. —? C. *goodenoughii* auct. non J. Gay: Forbes and Hemsley, Index Fl. Sin. 3 (1905) 287. —**Ic.**: Russk. Bot. zh. 3–6 (1911), fig. 72.

Described from Far East. Type in Leningrad.

In marshy sedge meadows and swamps.

IA. Mongolia. ***Cis Hing.*** (Yaksha railway station, in swamp, Nos. 2168, 2170 and 2184, 1954—Wang); ***Cent. Khalkha, Depr. Lakes, East. Mong.*** (near Khailar town, on wet site, No. 581, 8 VI; near Manchuria station, wet meadow, No. 669, 11 VI 1951, S.H. Li et al.; Khalkha-Gol somon, Salkhit hill region, Dege-Gol river floodplain, sedge marshy meadows, 10 VI; same site, Numurgin-Gol river region, sedge marshy mounds, 11 VI; Bayan-ula somon, sedge marshy meadows, 31 VII; same site, 7 km south of somon, valley floor, sedge marshy meadow, 31 VII 1956—Dashnyam).

General distribution: Arct. (Asiat., rare), West. Sib. (Altay), East. Sib. (predom. east of Lena river), Far East, Nor. Mong. (Hent., Mong-Daur.), China (Dunbei, ?North and East).

23. **C. cespitosa** L. Sp. pl. (1753) 978; Kük. in Pflanzenreich [IV, 20] 38 (1909) 328 (excl. var. *minuta*); Krylov, Fl. Zap. Sib. 3 (1929) 467, p.p.; V. Krecz. in Fl. SSSR, 3 (1935) 217; Grubov, Consp. fl. MNR (1955) 82; Fl. Kazakhst. 2 (1958) 56; Hanelt and Davažamc in Feddes Repert, 70, 1–3 (1965) 17. —C. *rubra* Lévl. and Vaniot in Bull. Ac. Intern. Géogr. Bot. 19 (1909) 33; V. Krecz. l.c. 218. —**Ic.**: Fl. SSSR, 3, Plate XV, fig. 6.

Described from Sweden. Type in London (Linn.).

In marshy sedge meadows.

IA. Mongolia: ***Khobd.*** ("Achit-nur"—Grubov, l.c.); ***Cis-Hing.*** (Khalkha-Gol somon, Khamar-Daban region, floodplain of river, 19 VI 1954; same site, region of Salkhit hills, floodplain of Dege-Gol river, sedge marshy meadow, 10 VI 1956—Dashnyam); ***Mong. Alt., East. Mong., Depr. Lakes*** (Shargin-Gobi, between Gol-ikhe and Tszakzbo, marshy hummocky coast of Shargin-Gol river, 6 IX 1930—Pob.); *West Gobi* ("Cagan-bulag, an der Quelle, No. 1048, VI 1962"—Hanelt and Davažamc, L.c.).

General distribution: Aral-Casp., Balkh. Region; Arct. (Europe; Asiat., rare), Europe, Caucasus (Nor., West. Transcaucasus), West. Sib., East. Sib., Far East, North Mong. (Fore Hubs., Hent., Hang.), China (Dunbei), Japan.

24. **C. ensifolia** (Turcz. ex Gorodk.) V. Krecz. in Fl. Zabaik. 2 (1931) 120 and in Fl. SSSR, 3 (1935) 227; Grubov, Consp. fl. MNR (1955) 84. —*C. rigida* ssp. *ensifolia* Turcz. ex Gorodk. in Zh. russk. bot. obshch. 15 (1930) 182. —*C. ensifolia* Turcz. ex Bess. in Flora, 17, 1, Beibl. (1834) 26 (nom. nud.). —*C. rigida* auct. non Good.: Kük. in Pflanzenreich [IV, 20], 38 (1909) 299, p.p.

Described from East. Siberia (Transbaikal). Type in Leningrad.

IA. Mongolia: ***Khobd.*** (Ulei-Daban, in forest belt, 14 VI 1879—Pot.); ***Mong. Alt.***

General distribution: Arct., Europe (Central Europ. hills, extreme nor.-east. Europ. SSSR and Urals), East. Sib., North Mong. (Fore Hubs., Hent., Hang.).

Note. The above cited specimen of *C. ensifolia* is not entirely typical; it tends towards *C. orbicularis* Boott.

25. **C. fuscovaginata** Kük. in Bull. Hérb. Boiss. 2 sér. 4 (1904) 56; idem, Pflanzenreich [IV, 20], 38 (1909) 338; Krylov, Fl. Zap. Sib. 3 (1929) 473; V. Krecz. in Fl. SSSR, 3 (1935) 211, p.p. (excl. pl. europ.); Grubov, Consp. fl. MNR (1955) 84; Fl. Kazakhst. 2 (1958) 56.

Described from Siberia (Altay). Type in Leningrad.

Along banks of water reservoirs, sometimes in water.

IIA. Junggar: ***Altay Region*** (Shara-Sume, 440 m, No. 236, 25 VIII; Qinhe, Chzhunkhaitsza, along fringe of water reservoir, in water, 2400 m, No. 1379, 6 VIII [evidently mixed with *C. orbicularis* Boott] 1956—Ching).

General distribution: Balkh. Region, Jung.-Tarb.; West. Sib. (south.), East. Sib., North Mong. (Hent., Hang).

Note. This is a critical species calling for special investigation. It is perhaps identical to *C. acuta* L. (*C. gracilis* Curt.). The only difference between the species is that the sheath of *C. fuscovaginata*, surrounding the shoot base, does not carry a leaf blade (i.e., the plant is an aphyllopod) while the sheath of *C. acuta* extends into a leaf blade (phyllopod). This feature is, very unstable, however.

26. **C. orbicularis** Boott in Trans. Linn. Soc. 20, 1 (1846) 134; Kük. in Pflanzenreich [IV, 20], 38 (1909) 303, p.p.; Pampanini, Fl. Carac. (1930) 84; V. Krecz. in Fl. SSSR, 3 (1935) 224; Persson in Bot. Notis. (1938) 276; Fl. Kirgiz. 2 (1950) 278; Grubov, Consp. fl. MNR (1955) 86; Fl. Kazakhst. 2 (1958) 57; Fl. Tadzh. 2 (1963) 104; Ikonnikov, Opred. rast. Pamira (Key to Plants of Pamir) (1963) 81. —*C. glauca* α. *typica* Rgl. and β. *brachylepis* Rgl. in Acta Hort. Petrop. 7, 2 (1880) 572. — *C. cespitosa* β. *vulgaris* and δ. *microstachys* Rgl. l.c. 574. —?*C. melanolepis* Boeck. Cyper. 1 XI (1888) 47; Fl. Kirgiz. 2 (1950) 281. —*C. erostrata* Boott ex Clarke in Hook. f. Fl. Brit. Ind. 6 (1873) 711; Strachey, Catal. (1906) 202. —*C. arcatica* Meinsh. in Acta Horti. Petrop. 18 (1901) 336 (incl. var. *pedunculata* Meinsh.); V. Krecz. l.c. 225; Fl. Kirgiz. 2 (1950) 281; Fl. Kazakhst. 2 (1958) 58; Fl. Tadzh. 2 (1963) 107. —*C. taldycola* Meinsh. l.c. 339; V. Krecz. l.c. 228; Fl. Kirgiz. 2 (1950)

281; Fl. Kazakhst. 2 (1958) 57. —*C. orbicularis* var. *brachylepis* (Rgl.) Kük. l.c. 304. —*C. orbicularis* var. *taldycola* (Meinsh.) Kük. l.c. 304. —*C. satakeana* T. Koyama in Acta Phytotax. and Geobot. 15, 4 (1954) 113. —*C. pakistanica* T. Koyama in Acta Phytotax. and Geobot. 17, 4 (1958) 100. —*C. rigida* auct. non Good.: Deasy, In Tibet a. Chin. Turk. (1901) 399; Hemsley, Fl. Tibet (1902) 201. —*C. cespitosa* auct. non L.: Walker in Contribs. U.S. Nat. Herb. 28 (1941) 599. —?*C. rigida* var. *dacica* auct. non Kük.: Pampanini, Agg. Fl. Carac. (1934) 156. —**Ic.:** Fl. Tadzh. 2, Plate XXII, figs. 5–8.

Described from Himalayas. Type in London (Linn.). Map 7.

In marshy and wet, sometimes solonetz-like meadows, in grass-sedge swamps, in marshy short-grass meadows along banks of rivers and brooks, near springs; predominantly in high-altitude belt, but descending to lower hill belt and even into foothills.

IA. Mongolia: ***Khobd.*** (Katu river, wet short-grass meadow, 16 VI; Altyn-Khatysyn, 18 VI; Bairimen-daban pass, 20 VI; valley of Kharkhiry river, on wet soil, 21 VII; Kharkhiry hill top, 24 VII 1879—Pot.); ***Mong. Alt.*** (Urten-Gol, meadow, 2 VII; Bura area, on swamp, 3 VII 1877—Pot.; Bodonchi river, hummocky bank, 19 VII 1898—Klem.; floodplain of Bodonchi river 2–3 km south of Bodonchin-khure, fringe of silted stream, 19 VII; Bulgan somon, upper Indertiin-Gol river, marshy meadow in high-altitude zone, 24 VII; same site, valley of Dundu-Tumurte river, midcourse, steppe meadow, 25 VII; same site, upper Ketsu-Sairin-Gol river, alpine meadow, 26 VII; Tamchi somon, 25–30 km south of Tamchi-daba pass, midcourse of Bidzhi-Gol river, birch grove near spring source, 10 VIII; same site, Bidzhi-Gol river, saline meadow with birch remnants near spring mouth along slope of plain, 10 VIII 1947—Yun.); ***Cent. Khalkha, Depr. Lakes, Val. Lakes, Gobi-Alt.*** (Dundu-Saikhan hills, 2 VII 1909—Czet.; Dzun-Saikhan hills, top of Yalo creek, wet meadow, 23 VII 1931—Ik.-Gal.; Dundu- and Dzun-Saikhan mountain ranges, along slopes, VII-VIII 1933—Simukova; Artsa Bogdo, No. 155, 1925—Chaney); ***East. Gobi*** (Dalan-Dzadagad town, rather saline meadows along brook, 1 km east of town, V-VII 1939—Surmazhab); ***West. Gobi*** (Tsagan-Bogdo mountain range, Tsagan-Bulak area, saline meadow near spring, 1 VIII; same site, Suchzhi-Bulak spring, rather saline meadow, 4 VIII 1943—Yun.); ***Alash. Gobi*** (Dyn-yuan'-in, along bed of brook, 3 VI 1908—Czet.: "Shui Mo Kou, Ho Lan Shan, No. 90; mouth of Hsi Yeh Kou, No. 183—1923, Ching"—Walker, l.c.); ***Khesi*** (valley of Khoyudung river, near Liyuan'-in town, 2 VI; Kheikho river between Gaotai and Fuiitin, in swamp, 6, 12 and 18 VI; Edzin-Gol river above Gaotai, 20 VI 1886—Pot.; Sachzhou, wet meadow, 1000 m, 14 VI 1894—Rob.).

IB. Kashgar: ***Nor.*** (Uchturfan, Kichik-Aran, in a garden, 14 V; same site, Kukurtuk gorge, 27 V 1908—Divn.; valley of Muzart river in Aksu district, No. 829, 9 IX 1958—Lee and Chu; Kucha, 2590 m, No. 10058, 26 VII 1959—Lee et al.); ***West.*** (12 km south of Upal oasis on Kashgar-Tashkurgan road, sasa zone, 11 VI 1959—Yun.; "Kashgar, 1330 m, 12 V 1935"—Persson, l.c.); ***South.*** (Maldzha [Niya] river, 12 VI 1885—Przew.; 20 km nor.-east of Niya settlement, on floodplain, Nos. 00046 and 00047, 7 VI 1959—Lee et al.; same site, saline meadow, 7 V; 10 km west of Keriya town, saline floodplain, low marshy sedge meadow, 10 V; Keriya oasis, 10 km north of town, Bostan area, sasa zone, marshy meadow, 14 VI; 8 km west of Keriya town, Chira-Khotan area, 16 V 1959—Yun; right bank of Keriya river, floodplain, No. 00062 and 00118, 10 and 16 VI 1959—Lee et al.); ***East.*** (Khami, on humus soil, in water, 29 V 1877—Pot.; Ledun in Khami, along bank of lake, No. 463, 21 V 1957—Kuan; in Turfan, near water, 100 m, No. 5507, 1 VI; Kurlya, alongside canal, 980 m, No. 5883, 14 VII 1958—Lee and Chu).

IC. Qaidam: ***Plain.*** (Ikhe- and Baga-Tsaidamin-Nor lakes, marsh, 3000 m, 14 VI 1895—Rob.).

IIA. Junggar: ***Altay Region*** (Qinhe, 2440 m, 4 VIII 1956—Ching); ***Jung. Alt.*** (Ven'tsyuan', No. 4634, 25 VIII 1957—Chin); ***Tien Shan, Jung. Gobi*** (Santakhu, 12 V 1877—Pot.; Sygashu,

600 m, 4 V 1879—A. Reg.; 2–4 km nor. of St. Kuitun settlement, on Shikho-Manas road, sasa zone, herb-sedge meadow fed by spring, 30 VI and 6 VII 1957—Yun.).

IIIA. Qinghai: ***Nanshan*** (Bardun river, 12 V 1886—Pot.; Sharagol'dzhin river, Bugu-Tologoi area, 14 VI; same site, Baga-Bulak, in marsh and in wet sand, 3150 m, 15 VI 1894—Rob.); ***Amdo*** (Huang He river upper course [Dzurge-Gol] river, 12 IV 1880—Przew.).

IIIB. Tibet: Chang Tang (Muzlyk-atasy mountain range along Muzlyk river, along springs, 3300 m, 16 VIII 1890—Rob.; "34°55' N lat., 82°15' E long., 4890 m"—Deasy, l.c.); ***Weitzan*** (valley of Alyk-Norin-Gol river, Kuku-Bulak spring, marshy banks, 3750 and 3600 m, 4 and 7 VI; basin of Yangtze river, Rkhombo-mpo lake, along Ichu river, 3930 m, 4 VIII 1900—Lad.); ***South.*** (Khambajong, No. 149, 26 VII 1903—Younghusband; "circa Shigatse, 28 VI 1914, No. 104195, E. Kawaguchi"—Koyama, l.c.).

IIIC. Pamir (Chumbus river, Kosh-Terek area, near water, 6 and 7 VI; Kenkol river, Tom-Kara area, marshy hummocky meadow, 14 VI 1909—Divn.; midcourse of Kanlyk river, 2500 to 3200 m, 12 VII: near Takhtakorum pass, 4500–5500 m, 1 VIII 1942—Serp.; valley of Sarykol river, 25 km south of Bulunkul' settlement, on Kashgar-Tashkurgan highway, marshy rather saline meadows, 12 VI 1959—Yun.).

General distribution: Balkh. Region, Jung.-Tarb., Nor. and Cent. Tien Shan, East. Pam.; Near East (Afghanistan, Pakistan—Hindukush), Mid. Asia (West. Tien Shan, Pamiroalay), West. Sib. (Altay), North Mong. (Fore Hubs., Hent., Hang. Mong-Daur.), China (Altay, North-west., South-west.), Himalayas (west., Kashmir).

Note. *C. orbicularis*, a characteristic species of Central Asia, is extensively distributed within its boundaries. This plant is quite variable in almost all of its characteristics but mainly in the height of stems, length of inflorescence, length of spikelets and the extent of their spacing, colour of glumes and the form of their tips (from obtuse to acute) and in the length ratio of glumes and utricles.

Among *C. orbicularis*, plants are commonly (10) 15–50 cm tall, with (2) 2.5–5 cm long inflorescence, 1–1.5 (2) cm long pistillate spikelet, with sessile staminate and pistillate spikelets, or, less often, on 0.5–1 cm long peduncles. In Tien Shan (Soviet and Chinese) and Pamiroalay, along with specimens of common *C. orbicularis*, comparatively large specimens up to 70 cm tall, with inflorescence 5–8 (10) cm long, somewhat more elongated 2–2.5 (3–4) cm long pistillate spikelets, with lower portion (as well as staminate spikelet) mostly on up to 1.5 (2) cm long peduncles.

Plants of the first type are essentially confined to high altitudes although they are found in all mountain belts, including submontane plains. Plants of the second type inhabit predominantly the moderate and lower mountain zones but quite often grow even at high altitudes—up to 3500 m or higher. Both types have several transitional forms.

V. Kreczetowicz in "Flora of the USSR" placed the very large specimens of *C. orbicularis* among *C. arcatica* Meinsh. described from Kirgizia. But, as shown by a study of its type specimens, *C. arcatica* is indistinguishable from the type form of *C. orbicularis*. It may be pointed out that Meinshausen compared his species with *C. rigida*, from which it actually differs, but not with *C. orbicularis*. Plants regarded as *C. arcatica* by V. Kreczetowicz, in our view, cannot be regarded as a species. They evidently represent a form or variant of *C. orbicularis*.

One cannot also accept the independent status of species *C. taldycola* Meinsh. described from Chinese Tien Shan. The original author distinguished it from *C. cespitosa* L. Type specimens of *C. taldycola* have pale brown glumes of staminate spikelets. They exhibit no other differences from *C. orbicularis*. Plants like typical ones are encountered from time to time throughout the distribution range of *C. orbicularis* and represent most probably a form of the latter.

C. pakistanica T. Koyama and *C. satakeana* T. Koyama are regarded here as synonyms of *C. orbicularis*. The former was described from Pakistan (Swat Himalayas) and the latter from Tibet (Shigatse). *C. pakistanica*, known in fact only from one site, has been differentiated by the original author from the other members of group *Rigidae* (which also includes *C. orbicularis*) in that its pistillate spikelets bear staminate flowers at the top. An examination of a large number of specimens showed that this feature is also characteristic of some specimens of other species of group *Rigidae* although to a much lesser extent than in the members of other groups of section *Acutae*. There is no basis, therefore, to regard plants possessing the aforesaid feature as a distinct species. In all other characteristics, C. *pakistanica* is identical to C. *orbicularis*.

C. *satakeana* Koyama has been distinguished by the original author from the species of group *Rigidae* by the absence of a creeping rhizome. The description of this species corresponds accurately with that of C. *orbicularis*. This is not contradicted even by the characteristics of the rhizome of C. *satakeana* since it varies in C. *orbicularis*, depending on the conditions of growth, from moderately creeping to contracted and the same mat may hold shoots with creeping as well as shortened rhizomes. Sometimes specimens of C. *orbicularis* form tufts (judging from notes on labels). As a rule, an elongated creeping rhizone is not characteristic of C. *orbicularis* and this feature distinguishes it from other species of the group *Rigidae* (C. *ensifolia*, for example), among which almost every shoot grows in the mat horizontally for some length, i.e., exhibits a distinct creeping rhizome.

C. *orbicularis* is very close to C. *altaica* (Gorodk.) V. Krecz. and the two have transitional forms between them. C. *orbicularis* exhibits close affinity also with C. *ensifolia*.

27. **C. schmidtii** Meinsh. in Baer and Helmers. in Beitr. Pfl. Russ. Reich. 26 (1871) 224; Kük. in Pflanzenreich [IV, 20] 38 (1909) 326; V. Krecz. in Fl. SSSR, 3 (1935) 223; Kitag. Lin. Fl. Mansh. (1939) 111; Grubov, Consp. fl. MNR (1955) 88. —**Ic.:** Pflanzenreich [IV, 20], 38, fig. 50.

Described from the Far East. Type in Leningrad.

In wet and marshy meadows.

IA. Mongolia: *Cis-Hing.* (near Yaksha railway station, wet meadow, 13 VI 1902—Litw.); *East. Mong.*

General distribution: Arct. (Asiat.—east., very rare), East. Sib., Far East, North Mong. (Hent., Hang., Mong.-Daur.), China (Dunbei), Korea (nor.), Japan.

Section Atratae Fries ex Pax

28. **C. caucasica** Stev. in Mém. Soc. natur. Moscou, 4 (1813) 108; V. Krecz. in Fl. SSSR, 3 (1935) 252; Fl. Kirgiz. 2 (1950) 281; Fl. Kazakhst. 2 (1958) 59. —C. *atrata* ssp. *caucasica* Kük. in Pflanzenreich [IV, 20], 38 (1909) 400, p.p., excl. pl. Kashmir.

Described from Caucasus. Type in Leningrad.

In subalpine meadows.

IIA. Junggar: *Tien Shan* (Agyaz river 2100–2400 m, 26 VI 1878—A. Reg.; Nilki brook on Kash river, 2100 m, 8 VI; Aryslyn 2400 m, 10 VIII 1879—A. Reg.; Danu river, No. 1431, 17 VII; 8 km south of Nyutsyuan'tsza, alpine meadow, No. 4674, 17 VII; 10 km north of Chzhaosu, on slope, 1800 m, No. 3318, 15 VIII 1957—Kuan; Tsanma valley, left tributary of Kunges, subalpine herb-cereal grass meadow, 7 VIII: nor. foothill of Narat mountain range descending into Tsanma valley, subalpine grass meadow, 7 VIII 1958—Yun.).

General distribution: Jung.-Tarb., North and Centr. Tien Shan; Europe (South. Urals), Asia Minor, Near East (Iran), Caucasus, Mid. Asia (West. Tien Shan).

29. **C. duthiei** Clarke in Hook. f. Fl. Brit. Ind. 6 (1893) 731. —C. *atrata* var. *pullata* Boott, Illustr. Carex 3 (1862) 114 and 115. —C. *atrata* ssp. *pullata* (Boott) Kük. in Pflanzenreich [IV, 20], 38 (1909) 400; Rehder in J. Arn. Arb. 14 (1933) 5; Walker in Contribs. U.S. Nat. Herb. 28 (1941) 599;

T. Koyama in J. Jap. Bot. 30, 10 (1955) 313. —*C. atrata* auct. non L.: Walker, l.c. 599. —**Ic.:** Boott, l.c. Tab. 364 (sub *C. atrata* var. *pullata*).

Described from the Himalayas. Type in London (K). Plate IV, fig. 3.

IIIA. Qinghai. ***Nanshan*** ("Tai Hua, No. 528; Ching Kang Yai, No. 576 [?]; La Chi Tzu Shan, No. 689–1923, R.C. Ching"—Walker, l.c.); ***Amdo*** ("alpine region between Radja and Jupar ranges, Rock"—Rehder, l.c.).

General distribution: China (North-west., South-west., Taiwan), Himalayas (west., east.), Indo-Malay. (Burma).

30. **C. hancockiana** Maxim. in Bull. Soc. natur. Moscou, 54, 1 (1870) 66; Forbes and Hemsley, Index Fl. Sin 3 (1905) 288; Kük. in Pflanzenreich [IV, 20], 38 (1909) 395; V. Krecz. in Fl. SSSR, 3 (1935) 271; Kitag. Lin. Fl. Mansh. (1939) 102; Grubov, Consp. fl. MNR (1955) 84.

Described from Nor. China. Type in Leningrad.

On wet sites in forest belt.

IA. Mongolia: ***Cis-Hing., Alash. Gobi*** (cent. and west. parts of Alashan mountain range, in marshy gorge, 5 VII 1873—Przew.).

IIA. Junggar: ***Tien Shan*** (Danu river, on wet sites, No. 1462, 17 VII 1957—Kuan; Manas river basin, valley of Koisu river as it enters Ulan-su river, valley floor, grass plot fed by stream, 17 VII 1957—Yun.).

General distribution: West. Sib. (Altay), East. Sib. (Baikal Region), North Mong. (Hang.), China (Dunbei, North, North-west.), Korea.

31. **C. infuscata** Nees in Wight. Contribs. Bot. Ind. (1834) 125 (excl. var. β.), emend. Raymond in Biol. Skr. Dan. Vid. Selsk. 14, 4 (1965) 26. —*C. alpina* var. *infuscata* (Nees) Boott, Ill. Carex, 3 (1862) 113, pro max. p.; Danguy in Bull. Mus. nat. hist. natur. 17, 6 (1911) 451. —? *C. atrofurfur* T. Koyama in Acta Phytotax. and Geobot. 16, 6 (1956) 166. —**Ic.:** Boott, l.c., Tab. 358.

Described from Nepal. Type in London (K).

IIIA. Qinghai: ***Nanshan*** (between Nanshan and Donkyru mountain ranges along Rakogol river, marshy sites among shrubs, 3000–3300 m [21 VII] 1880—Przew.; "Jong-Ngan, alt. 3200 m, 9 VII 1908, L Vaillant"—Danguy, l.c.).

General distribution: Near East (Afghanistan, Pakistan), Himalayas.

Note: *C. atrofurfur* T. Koyama, described from southern Karakorum (Oltali Chish), is evidently identical to *C. infuscata*. The original author compares *C. atrofurfur* with *C. alpina* Sw. ex Liljebl. (= *C. norvegica* Retz.), but the characteristics on the basis of which he differentiated his species from the latter, also distinguish *C. infuscata* from it.

The sedge cited by V. Kreczetowicz in "Flora of the USSR" under the name *C. infuscata* was later established by him as a new species, *C. popovii* V. Krecz.

32. **C. kansuensis** Nelmes in Kew Bull. (1939) 201. —? *C. atrata* auct. non L.: Forbes and Hemsley, Index Fl. Sin. 3 (1905) 274. —*C. atrata* var. *ovata* auct. non Boott: Danguy in Bull. Mus. nat. hist. natur. 17, 6 (1911) 451.

Described from North-west. China (Gansu province). Type in London (K). Plate I, fig. 4.

In swampy meadows in alpine and moderate mountain belts.

IIIA. Qinghai: ***Nanshan*** (South.-Tetungsk mountain range, on marshy meadow in alpine belt, 7 IX 1872—Przew.; Sinin hills, Myndan'sha river, 29 V 1890—Gr.-Grzh.; Kukunor lake, Uiyu area, south. slope of central belt, 13 VIII 1908—Czet.; "Jong-tchoun, alt. 3400 m, 6 VII 1908, L. Vaillant"—Danguy, l.c.); ***Amdo*** (hills in valley of Mudzhik river, 3150–3450 m, 22 VI 1880—Przew.).

General distribution: China (North-west., ?Cent., South-west.).

Note. The species is closer to C. *perfusca* V. Krecz., differing mainly in the presence of sparse ciliate spinules along margins of utricles in upper part as well as in lighter coloured, brownish-mottled, but not uniformly coloured utricles.

33. **C. lehmannii** Drej. Symb. Caric. (1844) 13; Forbes and Hemsley Index Fl. Sin. 3 (1905) 293; Kük. in Pflanzenreich [IV, 20], 38 (1909) 387. —**Ic.:** Drejer l.c. Tab. II.

Described from Nepal. Type in London (K). Plate V, fig. 3.

In forest belt.

IIIA. Qinghai: ***Nanshan*** (South.-Tetungsk mountain range, in lower forest belt, 11 VIII 1872; same site, in central forest zone, in forest, 2550 m, 3 VIII; same site, in lower forest zone, 2250 m, 7 VIII 1880—Przew.).

General distribution: China (North-west., Centr. and South-west.), Himalayas.

34. **C. melanantha** C. A. Mey. in Ledeb. Fl. Alt. 4 (1833) 216; Duthie in Alcock, Rep. Pamir Commiss. (1898) 27; Kük. in Pflanzenreich [IV, 20] 38 (1909) 391, p.p.; Krylov, Fl. Zap. Sib. 3 (1929) 477; Pampanini, Fl. Carac. (1930) 84; V. Krecz. in Fl. SSSR, 3 (1935) 273; Persson in Bot. Notis. (1938) 276; Fl. Kirgiz. 2 (1950) 286; Grubov, Consp. fl. MNR (1955) 86; Fl. Kazakhst. 2 (1958) 60; Fl. Tadzh. 2 (1963) 110. —C. *melanantha* var. *moorcroftii* (Boott) Kük. l.c. 391, quoad pl. turkest. —**Ic.:** Fl. SSSR, 3, Plate XVI, fig. 5.

Described from Altay. Type in Leningrad. Plate II, fig. 2; map 8.

In wet and marshy meadows, on hill steppes, along banks of rivers and springs; from alpine belt with very extensive distribution up to upper forest boundary.

IA. Mongolia: ***Khobd.*** ("Achit-Nur"—Grubov, l.c.); ***Mong. Alt.*** (upper Naryn river, on slope in larch forest, 23 VII 1898—Klem.; Chara-Khargai—Kharga pass, alpine steppe, 10 VIII 1909—Sap.; Adzhi-Bogdo mountain range, ascent to Burgastyin-daba pass in upper Indertiin-Gol, hill steppe, 6 VIII; Bulugun somon, Kharagaitu-daba pass, alpine meadow, 24 VII; Adzhi-Bogdo mountain range, central part of Burgastyin-daba pass, hill steppe, 6 VIII 1947—Yun.); ***East. Mong.***

IIA. Junggar: ***Altay Region*** (Qinhe district, Chzhunkhaitsza settlement, 4 VIII 1957—Kuan); ***Jung Alt.*** (30 km west of Ven'tsyuan', No. 2024, 25 VIII 1957—Kuan), Tien Shan.

IIIA. Qinghai: ***Nanshan*** (Humboldt mountain range [Argalin-ula], short-grass meadow near brook on sand, 3600–4200 m, 20 VI; Nanshan mountain range, 3450 m, 11 VII 1894—Rob.).

IIIC. Pamir (Billuli river as it flows into Chumbus, 12 VI 1909—Divn.; "Fairly common along the banks of the Aksu river and its affluents, 3900–4200 m, Nos. 17771 and 17772, Alcock"—Duthie, l.c.; "Pamir, Mintaka, about 4150 m, 3 VII 1935"—Persson, l.c.).

General distribution: Jung.-Tarb., North and Cent. Tien Shan, East. Pamir; Near East (Afghanistan, Pakistan—Chitral), Mid. Asia (West. Tien Shan and Pamiroalay), West. Sib. (Altay), East. Sib. (East. Sayan mountain range), North Mong. (Fore Hubs., Hang.), Himalayas (west., Kashmir).

35. **C. melananthiformis** Litv. in Tr. Bot. muz. AN, 7 (1910) 90; V. Krecz. in Fl. SSSR, 3 (1935) 274; Fl. Kirgiz. 2 (1950) 286; Grubov, Consp. fl. MNR (1955) 86; Fl. Kazakhst. 2 (1958) 62. —*C. melanantha* auct. non C.A. Mey.: Kük. in Pflanzenreich [IV, 20], 38 (1909) 391, p.p., quoad pl. sibir. —*C. melanantha* var. *moorcroftii* (Boott) Kük. l.c. 391, quoad pl. sibir. —**Ic.:** Fl. SSSR, 3, Plate XVI, fig. 6.

Described from East. Siberia (East. Sayan mountain range). Type in Leningrad. Plate II, fig. 1.

In marshy meadows, sedge marshes, on steppe slopes, along banks of rivers; from alpine to upper forest belt.

IA. Mongolia: ***Khobd.*** (Ikki-daba, 21 VI 1870—Kalning; Altyn-Khatysyn area, on dry sandy soil, 19 VI; Tszusylan, in forest, 13 VII 1879—Pot.); ***Mong. Alt.*** (10 km south-east of Yusun-Bulak, Khan-Taishiri mountain range, herb-wheat grass-chee grass steppe, 14 VII 1947—Yun.); ***Gobi-Alt.*** (Dzun-Saikhan hills, on meadow, on top of Yalo creek, 23 VIII 1931—Ik.-Gal.); ***East. Gobi*** (Alashan-Urgu Road, Urten-Gola valley, 22 VI [1909]—Czet.).

IIA. Junggar: ***Tien Shan*** (B. Yuldus river, in swamp, in water, 2460 m, No. 6468, 10 VIII 1958—Lee and Chu; B. Yuldus basin, 30–35 km south-west of Bain-Bulak settlement, grassy sedge marsh, 10 VIII 1958—Yun.).

General distribution: Balkh. Region, Jung.-Tarb., Centr. Tien Shan; Caucasus, West. Sib. (Altay), East. Sib. (south.), North Mong. (Fore Hubs., Hent.).

Note. Plants with staminate upper spikelet are found from time to time among *C. melananthiformis* Litv.

36. **C. melanocephala** Turcz. in Bull. Soc. natur. Moscou, 28, 1 (1855) 334; ej. Fl. baic.-dahur. 2, 1 (1856) 269; V. Krecz. in Fl. SSSR, 3 (1935) 266; Fl. Kirgiz. 2 (1950; Grubov, Consp. fl. MNR (1955) 86; Fl. Kazakhst. 2 (1958) 60. —«*C. parviflora* C. A. Mey»: Kük. in Pflanzenreich [IV, 20], 38 [1909] 386, p.p.; Krylov, Fl. Zap. Sib. 3 (1929) 476. —**Ic.:** Fl. SSSR, 3 Plate XVI, fig. 7.

Described from East. Siberia (East. Sayan mountain range). Type in Leningrad.

In alpine meadows.

IA. Mongolia: ***Mong. Alt.*** (Oi-Chilik alpine zone, 20 IX 1876—Pot.; Bulugun somon, Kharagaitu-daba pass in Indertiin-Gola upper courses, alpine meadow, 24 VII 1947—Yun.).

IIA. Junggar: ***Altay Region*** (between Qinhe and Tsaganhe rivers, 3370 m, No. 1491, 8 VIII 1956—Ching); ***Tien Shan*** (Aryslyn, 2700–3000 m, VII 1879—A. Reg.).

General distribution: Jung.-Tarb., Cent. Tien Shan, West. Sib. (Altay), East. Sib. (south.), North Mong. (Fore Hubs., Hang.), China (Altay).

*__C. meyeriana__ Kunth, Enum. pl. 2 (1837) 438; Kük. in Pflanzenreich [IV, 20], 38 (1909) 393; V. Krecz. in Fl. SSSR, 3 (1935) 262; Kitag. Lin. Fl. Mansh. (1939) 106; Grubov, Consp. fl. MNR (1955) 86.

Described from East. Siberia. Isotype in Leningrad.

In marshes and marshy meadows.

Found on Fore Hing. border of Mongolian region (near Irekte railway station, on bog, 16 VI 1902—Litw.).

General distribution: West. Sib. (Altay), East. Sib., Far East (south.), North Mong. (Fore Hubs., Hent., Hang.), China (Dunbei), Korea, Japan.

37. **C. moorcroftii** Falc. ex Boott in Trans. Linn. Soc. 20, 1 (1846) 140 and Illustr. Carex, 1 (1858) 9; Henderson and Hume, Lahore to Yarkand (1873) 338; Hemsley in J. Linn. Soc. (London) Bot. 30 (1894) 119; ej. Fl. Tibet (1902) 201; Deasy, In Tibet and China. Turk. (1901) 404; Strachey, Catal. (1906) 202; Keissler in Ann. Naturhist. Hofmus. 22 (1907–1908) 32; Ostenfeld in Hedin, S. Tibet, 6, 3 (1922) 91; Hand.-Mazz. in Österr. Bot. Z. 79 (1930) 39. —*C. melanantha* var. *moorcroftii* (Boott) Kük. in Pflanzenreich [IV, 20], 38 (1909) 391, p.p.; Pampanini, Fl. Carac. (1930) 84. —*C. sabulosa* auct. non. C. A. Mey.: V. Krecz. in Fl. SSSR, 3 (1935) 275, p.p., quoad pl. tibet. —Ic.: Boott, l.c. 1858, Tab. 27.

Described from the Himalayas. Type in London (K). Plate II, fig. 3; map 8.

On fixed and loose sand, in sandy meadows, on sandy and rocky banks of rivers in high-altitude belt. Abundant and a valuable fodder plant in the high-altitude pastures of Tibet; it is also a loose-sand binder.

IC. Qaidam: Mount. (Baga-Qaidam in-Nor lake, in sandy meadow, 3000 m, 5–8 VI 1895—Rob.).

IIIA. Qinghai: ***Nanshan.*** (Humboldt mountain range, Kuku-usu river, 6 VI 1894—Rob.).

IIIB. Tibet: ***Chang Tang*** (Prov. Yárkand, plateau at foot of the Karakorúm, north-east of the Karakorúm Pass, 10–11 VIII; Prov. Khótan, Lake Kiük-Kiöl, 13–14 VIII 1856, Schlagintweit; Muzlyk-atasy hills, in wet sand, 3900 m, 16 VIII; Przewalsky mountain range, nor. slope, in sand, 4200–4500 m, 22 VIII 1890—Rob.; Hattan-ning-davan, 5400 m, 19 VIII 1892—Rhins; basin of Karakash river, upper Khotan-Darya river, 10–12 km west of Shakhidulla in Kirgiz-Daban pass, short-grass meadow on sandy alluvium, 3900 m, 3 VI 1956—Yun.; "from Chágra to the Karakash Valley, 4500–5400 m, 1870"—Henderson and Hume, l.c.; "Tibet, sandy gravelly soil, at 5280 m, Thorold"—Hemsley, l.c. 1894, 1902; "hillslope two miles north of Murusu river, 35° 53′ N lat., 90°31′ E. long., 21 VI 1892, Rockhill"—Hemsley, l.c. 1902; "Northern Tibet, Shor Kul, 4500 m, 1897–1899"—Deasy, l.c.; "Ullug Kul, 4950 m, 26 VI; Quellfluß des Keria Darja, 5200 m, 6 VII 1906, Zugmayer"—Handel-Mazzetti, l.c. and Keissler, l.c.; Valle del Caracàsh sup., presso Chisil-Gilga, 5100 m, No. 8, Dainelli e Marinelli"—Pampanini, l.c.; "Lago Páncong, fra il lago e Mógaleb, 4350 m, Roero"—Pampanini, l.c.; "Valle del Ciángcemmo, Cone-la, Rhins"—Pampanini, l.c.; "Northern Tibet, Mandarlik, 3437 m, VII 1900, S. Hedin"—Ostenfeld, l.c.): ***Weitzan*** (left bank of Dychu river, 4590–3810 m, 21 VI; between Huang He and Yangtze rivers, 12 VI 1884—Przew.; Burkhan-Budda mountain range, nor. slope, Nomokhun gorge, on sandy-clayey soil, 17 V; basin of Huang He river, Russkoe lake and Serchu river, on sandy-ricky banks, 4050 m, 27 V; Alyk-Norin-Gol, 3600 m, No. 125a, undated; Amnen-kor mountain range, south. slope, 3900–4200 m, 9 VI; Dzhagyn-Gol river, 3750 m, 2 VII 1900—Lad.); ***South.*** (Khambajong, 8–10 VII 1903—Younghusband; same site, IX 1903—Prain).

General distribution: Himalayas (west., Kashmir).

Note. Most characteristic plant of sandy Tibetan uplands. The species is very close to North Mongolian-East. Siberian *C. sabulosa* Turcz. ex Kunth. Unlike the latter, the apical spikelet in *C. moorcroftii* is staminate (less frequently, androgynous), leaves very broad, 3–5 (6) mm broad, straight, erect, strongly short-acuminate and mostly flat. Also utricles in *C. moorcroftii* are very small, 3.5–4 mm long, with very short beaks. [In *C. sabulosa,* the apical spikelet is usually gynaecandrous, leaves 2–3 (4) mm long, often longitudinally folded or with recurved margin, generally curved or sinuate, long-acuminate, utricles (4) 4.5–5 (6) mm long.] The difference between these species in the structure of the apical spikelet is not absolute. Within the distribution range of C. *sabulosa*, plants with staminate upper spikelet are occasionally found while specimens with gynaecandrous upper spikelet are found, albeit very rarely, among C. *moorcroftii*.

Some specimens of C. *moorcroftii* (with broad flat leaves) are indistingushable from C. *melanantha* C. A. Mey. in external shape. The two species can be easily differentiated from utricles which are large with a bicuspid beak in the former and very small with entire beak in the latter.

38. **C. norvegica** Retz. Fl. Scand. (1779) 179. —C. *alpina* Sw. in Liljebl. Svensk. Fl. ed. 2 (1798) 26; Kük. in Pflanzenreich [IV, 20], 38 (1909) 384, p.p.; Krylov, Fl. Zap. Sib. 3 (1929) 475, p.p. —C. *mimula* V. Krecz. in Fl. SSSR, 3 (1935) 603 and 266. —C. *halleri* auct. non Gunn.: V. Krecz. l.c. 265; Egorova in Bot. mater. Gerb. Bot. inst. AN SSSR, 19 (1959) 80. —C. *angarae* auct. non Steud.: Grubov, Consp. fl. MNR (1955) 253, p.p. —**Ic.:** Pflanzenreich [IV, 20], 38 fig. 60 (A-E).

Described from Norway. Type?

IA. Mongolia: ***Khobd.*** (Tszusylan west of Ulangom, in forest, 13 VII 1879—Pot.).

General distribution: Arct., Europe (nor.), West. Sib. (Altay), East. Sib., Far East (Zeya basin), North Mong. Fore Hubs., Hang., Mong.-Daur.), Nor. Amer. (east.).

39. **C. perfusca** V. Krecz. in Fl. SSSR, 3 (1935) 600 and 256; Fl. Kirgiz. 2 (1950) 282; Grubov, Consp. fl. MNR (1955) 87; Fl. Kazakhst. 2 (1958) 59. —C. *atrata* var. *aterrima* auct. non Hartm.: Kük. in Pflanzenreich [IV, 20], 38 (1909) 398, p.p., quoad pl. sibir.; Krylov, Fl. Zap. Sib. 3 (1929) 480. —C. *atrata* auct. non L.: Krylov, l.c. 480, p.p., excl. pl. ural.; Kitag. Lin. Fl. Mansh. (1939) 97. — **Ic.:** Fl. SSSR, 3, Plate XVII, fig. 7.

Described from West. Siberia (Altay). Type in Leningrad.

In meadows and on meadow slopes in subalpine and alpine zones and in upper forest belt.

I.A. Mongolia: ***Khobd., Mong. Alt.***

IIA. Junggar: ***Jung. Alt.*** (17 km nor.-west of Ven'tsyuan', shaded slope, 2700 m, No. 1660, 29 VIII 1957—Kuan); ***Tien Shan*** (Bogdo hill, 2700 m, 24 VII 1878—A. Reg.; Karagol, 3000 m, 14 VI; south. tributary of Karagol, 2700 m, 16 VI; west. tributary of Aryslyn, 2400–2700 m, 8 and 28 VII; Aryslyn, 2400–2700 m, 10 and 12 VII; same site, 3000–3300 m, 13 VII 1879—A. Reg.; 3 km south of Yakou, shaded slope, 2900 m, Nos. 1664 and 1685, 30 VIII; between Ulastai and Yakou, on slope, No. 3969, 30 VIII 1957—Kuan)..

General distribution: Jung.-Tarb., North and Centr. Tien Shan, Arct. (Asiat.—lower Yenisey), Mid. Asia (West. Tien Shan), West. Sib. (South. Urals, Altay), East. Sib., Far East (Sea of Okhotsk, coast and south.), North Mong. (Fore Hubs., Hent., Hang.). Korea, Japan (Hondo island).

***C. pseudobicolor** Boeck. Cyper. nov. 1 (1888) 44, emend. Raymond in Biol. Skr. Dan. Vid. Selsk. 14, 4 (1965) 29. —C. *alpina* var. β. *erostrata* Boott, Illustr. Carex, 1 (1858) 71; Strachey, Catal. (1906) 202. —C. *alpina* ssp. *infuscata* (Nees) Kük. var. *erostrata* (Boott) Kük. in Pflanzenreich [IV, 20], 38 (1909) 386. —C. *infuscata* var. *erostrata* (Boott) T. Koyama in Acta Phytotax. and Geobot. 17, 4 (1958) 97. —**Ic.:** Boott, l.c. Tab. 194, fig. 2.

Described from the Himalayas. Type?

IIIC. Pamir. ("Pamir Highland"—T. Koyama, l.c.).

General distribution: Near East (Pakistan—Chitral), Himalayas (west.).

Note. The reported occurrence of this species in Pamir by Koyama (l.c.) is in all probability, erroneous.

40. **C. sabulosa** Turcz. ex Kunth, Enum. pl. 2 (1837) 432; Hemsley and Pearson in Peterm. Mitteil. 28, Heft 131 (1900) 375; V. Krecz. in Fl. SSSR, 3 (1935) 275, p.p., excl. pl. tibet.; Grubov, Consp. fl. MNR (1955) 88; Fl. Kazakhst. 2 (1958) 63, p.p., excl. pl. tibet. —*C. melanantha* var. *sabulosa* (Turcz. ex Kunth) Kük. in Pflanzenreich [IV, 20], 38 (1909) 392; Krylov in Fl. Zap. Sib. 3 (1929) 478. —**Ic.:** Fl. SSSR, 3 Plate XVI, fig. 4.

Described from East. Siberia (Baikal lake). Type in Leningrad.

Along banks on sands and dunes, sand steppes, dry meadows.

IA. Mongolia: ***Mong. Alt.*** (Kak-kul' lake, between Tsagan-Gol and Kobdo rivers, dry rubble meadow, 22 VI 1906—Sap), ***Centr. Khalkha*** (72 km south of Ulan-Bator, floor of hummocky valley, meadow steppe, 7 VIII 1951—Kal.), ***East. Mong., Depr. Lakes, Val. Lakes*** (40 km south of Ulyasutai on Tsagan-Ol road, sands, 17 VII 1947—Yun.).

General distribution: Balkh. Region (east.); Arct. (Asiat.—Lena lower courses), West. Sib. (?Altay), East. Sib., North Mong.

Note. The reported occurrence of this species in Tibet (Hemsley and Pearson, l.c.) pertains, in all probability, to *C. moorcroftii.*

C. musartiana Kük. ex V. Krecz. [Flora of the USSR, 3 (1935) 602, 261] described from Chinese Tien Shan on the basis of a few specimens collected by A. Regel (Musart, VIII, 1877, A. Regel) pertains to section *Atratae.* This plant was found to be identical to the East. Siberian-Circumpacific species *C. podocarpa* R. Br. established by V. Kreczetowicz himself (Bot. mater. Gerb. Bot. inst. AN SSSR, 9, 1941), with which view we dissociate altogether. The report that *C. podocarpa* (= *C. musartiana*) grows in Tien Shan is extremely doubtful, given the nature of the distribution range of this species (its western border falls in Baikal Region). Moreover, the occurrence of this species in Tien Shan has not been confirmed by the latter herbarium collections. In all probability, the specimens described as *C. musartiana* got mixed in Regel's collection and were improperly labelled.

Section Montanae (Fries) Christ

C. amgunensis Fr. Schmidt in Mém. Ac. Sci. St. Pétersb. sér. 7, 12 (1868) 69; Kük. in Pflanzenreich [IV, 20], 38 (1909) 447; Krylov, Fl. Zap. Sib. 3 (1929) 487; V. Krecz. in Fl. SSSR 3 (1935) 319; Kitag. Lin. Fl. Mansh. (1939) 97; Grubov, Consp. fl. MNR (1955) 81. —*C. chloroleuca* Meinsh. in Bot. Centralbl. 55 (1893) 196. —*C. amgunensis* var. *chloroleuca* (Meinsh.) Kük. l.c. 447. —**Ic.:** Fr. Schmidt, l.c. Tab. 1, figs. 4 and 5.

Described from the Far East (Amgun' river). Type in Leningrad.

In dry meadows and dry forests.

IA. Mongolia: ***Mong. Alt.*** (near Ulangom). Found also in the border region of Hinggan in Mongolia (near Irekte railway station, forest slope, 15 VI; same site, dry meadow, 16 VI 1902, Litw.).

General distribution: Europe (Cent. Urals), West. Sib. (spurs of Kuznetsk Alatau and Altay), East. Sib. (south.), Far East (south.), North Mong. (Fore Hubs., Hent., Hang.), China (Dunbei).

Section Mitratae Kük.

41. **C. caryophyllea** Latourr. Chlor. Lugdun. (1785) 27; Kük. in Pflanzenreich [IV, 20], 38 (1909) 463. —*C. verna* Chaix ex Vill. Hist. pl. Dauph. 1 (1786) 312; Krylov, Fl. Zap. Sib. 3 (1929) 490. —*C. ruthenica* V. Krecz. in Fl. SSSR, 3 (1935) 610, 325; Grubov, Consp. fl. MNR (1955) 87; Fl. Kazakhst. 2 (1958) 68. —**Ic.:** Fl. SSSR, 3, Plate XIX, fig. 8 (sub *C. ruthenica*).

Described from Europe (France). Type in Paris.

IA. Mongolia: ***Mong. Alt., Gobi-Alt.*** (Artsa Bogdo, hummock on stream edge at 1290 m, No. 159, 1925—Chaney).

General distribution: Aral-Casp., Balkh. Region; Europe, West. Siberia (south., excluding Altay), East. Siberia (south.), North Mong. (Fore Hubs., Hent., Hang., Mong.-Daur.).

42. **C. titovii** V. Krecz. in Tr. Sredneaz. univ., ser. 8-b, 20 (1935) 15; Fl. Kirgiz. 2 (1950) 291; Fl. Kazakhst. 2 (1958) 67.

Described from Junggar (Jung. Alatau). Type in Tashkent.

In steppe meadows in moderate hill zone.

IIA. Junggar: ***Jung. Alt.*** (Toli district, Bartok—Arba-kezen', in meadows, 1950 m, No. 1394, 8 VIII 1957—Kuan); ***Tien Shan*** (Ketmen' pass, 2400–2700 m, [19] VI; Khanakhai, [13–17] VI 1878—A. Reg.; Piluchi, [22] IV; Taldy, 2400–2700 m, 27 V; west. tributary of Aryslyn, 2400–2700 m, 8 VII 1879—A Reg.).

General distribution: Jung. Tarb., North and Centr. Tien Shan (rare).

Note. *C. titovii* is very close to the Siberian-European *C. caryophyllea* Latourr. (= *C. ruthenica* V. Krecz.), differing from it (according to V. Kreczetowicz) in acute and aristate glumes longer than utricles or as long, as well as in very large utricle 2.8–3 but not 2.3–2.5 mm. The herbarium of BIN (LE) contains not even one specimen of *C. titovii* conforming to V. Kreczetowicz's description; only on the basis of the site the above specimens were assigned to this species solely on the basis of the site of their collection. These specimens have elongated acute glumes but their utricles when fully mature are 2.5 mm long. Specimens with such glumes are found even among *C. caryophyllea*. The question of the independent specific status of Junggar and Tien Shan plants with affinity to *C. caryophyllea* has not been resolved thus far for want of adequate material.

Section Lamprochlaenae (Drej.) Kük.

43. **C. alba** Scop. Fl. Carn. ed. 2, 2 (1772) 216; Kük. in Pflanzenreich [IV, 20], 38 (1909) 500; Krylov, Fl. Zap. Sib. 3 (1929) 498; V. Krecz. in Fl. SSSR, 3 91935) 373; Fl. Kazakhst. 2 (1958) 70. —**Ic.:** Pflanzenreich [IV, 20], 38 fig. 79 (A–E).

Described from Austria. Type in Vienna (?).

IIA. Junggar: ***Jung. Alt.*** (Syaeda-Ven'tsyuan', 2700–2900 m, No. 1400, 13 VIII 1957—Kuan); ***Tien Shan*** (Fukan district west of Tyan'chi lake, nor.-east. slope, No. 1933, 19 IX 1957—Kuan).

General distribution: Jung.-Tarb. (Jung. Alatau); Europe (separately in West. Europe, nor. Europ. USSR, Cent. Urals), Caucasus (Dagestan), West. Siberia (south.—Saratov on Irtysh and Altay), East. Sib. (highly broken); North Mong. (Fore Hubs.).

44. **C. aridula** V. Krecz. in Bot. mater. Gerb. Bot. inst. AN SSSR, 9 (1946) 190.

Described from North-west. China. Type in Leningrad.

IA. Mongolia: ***Alash. Gobi*** (Alashan mountain range, Tsuburgan-Gol gorge, 29 IV 1908—Czet.).

IIIA. Qinghai: ***Nanshan*** (in valley of Lanchzha-lunva river, 12 V 1885—Pot., type!).

IIIB. Tibet: ***Weitzan*** (upper Huang He river, valley of Churmyn river, in marshy forest, 2700–2850 m, 14 V 1880—Przew.).

General distribution: China (North-west.—Gansu).

Note. The species has been little studied and all known specimens have altogether immature utricles. It is very close to *C. korshinskyi* Kom., from which it differs in broader, shorter and obtuse glumes.

45. **C. ivanoviae** Egor. in Nov. sist. vyssh. rast. (1966) 34.

Described from North-west. China. Type in Leningrad. Plate III, fig. 2; map 9.

On sand and sandy sites in high-altitude and moderate mountain zones (2200–4500 m above sea level).

IIIA. Qinghai: ***Nanshan*** (60 km south-east of Chzhan'e town, high foothills of Nanshan, 2200 m, 12 VII 1958—Petr.).

IIIB. Tibet: ***Chang Tang*** (Przewalsky mountain range, on sand, 4200–4500 m, 21 VIII 1890—Rob.: Cherchen district, 30 km south-east of Achan, on slope, in high-altitue belt, 4190 and 4260 m, Nos. 9413 and 9427, 4 VI 1959—Lee et al.): ***Weitzan*** (left bank of Dychu river [Yangtze], on sandy sites, 3900 m, 23 VI 1884—Przew., type!; valley of Alyk-Norin-Gol river, on dry sand, 3750 m, 3 VI 1900—Lad.; same site, on dry clayey-sandy sites, 3600 m, 8 VI 1901—Lad.).

General distribution: China (North-west.—Gansu province).

Note. Central Asian species allied to *C. turkestanica* Rgl., differing well from the latter in narrow leaves folded longitudinally and resembling bristles, brown leaf sheaths, one staminate spikelet and oblong-ovate utricles with glabrous beak.

While describing this species (Egorova, l.c.), an error occurred in the citation of the locality of the type specimen. Maximowicz's label, according to the citation read: left bank of Huang He river. However, according to the field label of Przewalsky and date of collection, this specimen was found on the left bank of the Dychu river (Yangtze river).

46. **C. korshinskyi** Kom. in Fl. Man'chzh. 1 (1901) 394; V. Krecz. in Fl. SSSR, 3 (1935) 375; Kitag. Lin. Fl. Mansh. (1939) 103; Grubov Consp. Fl. MNR (1955) 85. —*C. supina* var. *costata* Meinsh. in Acta Hort. Petrop. 18 (1901) 392; Kük. in Pflanzenreich [IV, 20], 38 (1909) 457. —*C. supina* var. *korshinskyi* (Kom.) Kük. l.c. 457.

Described from the Far East (South. Primor'e). Type in Leningrad. Map 9.

On rock and rubble steppe slopes, hummocky sands, placers, rocks; up to subalpine zone.

IA. Mongolia: ***Khobd.*** (upper Kharkhiry river, upland steppe, 5 IX 1931—Bar.); ***Cis-Hing.*** (Yaksha railway station, rocks, 11 VI 1902—Litw., same site, dry hill slope, No. 2192; same site, on rocks, No. 2254 (1954)—Wang; Khuna province, near Trekhrech'e, near marsh, 700 m, No. 1436, 16 VII 1951—Wang and S.H. Li; Khalkha-Gol somon, Dege-Gola region, south of Salkhit hill, southern slope, pea shrub-meadow grass-wheat grass steppe, 10 VI 1956—Dashnyam); ***Cent. Khalkha*** (Orgochen-sume monastery neighbourhood [47° N lat., 104° E long.], granite outcrops, 23 VIII 1925—Krasch. and Zam.; Bichikte-ulan-khada, south-east. slope, near rocks, 8 VII 1926—Lis.; granite outlier at Sorgol-Khairkhan, 180 km south-south-west of Ulan-Bator, along cracks in granite cliffs, 16 V; granite outlier at Choiren-ul on Ulan-Bator—Kalgan road, chee grass-wormwood steppe, 28 V; same site, among rocks, 7 VII

1941—Yun.; Mandal somon, Sakhil'te-ula, rock slopes, 20 V 1944—Yun.; Bayan-ula somon, 150 km north-west of Choibalsan, chee grass-tansy steppe, 31 VIII 1956—Dashnyam); ***East. Mong., Depr. Lakes*** (Ubsu-Nur lake, valley of Sagli river, between knolls, 4 X; sands near Khara-Bury, 4 X 1931, Bar.; Dzun-Gobi somon, Borig-del' sands south-east of Bain-Nur lake, hummocky sands with wormwood, 25 VII 1945—Yun.); ***Gobi-Alt.*** (Dundu-Saikhan hills, rubble placers on slope, 18 VIII; Dzun-Saikhan hills, on rubble slope, 25 VIII 1931—Ik.-Gal.; Ikhe-Bogdo mountain range, nor. slope, Tsagan-Burgas creek, hilly subalpine steppe, 15 IX 1943—Yun.).

IIA. Junggar: ***Altay Region*** (20 km nor.-west of Shara-Sume, scrub-meadow steppe, 7 VII 1959—Yun.); ***Tarb.*** (east. trail of Saur mountain range on Karamai-Altay road, sheep's fescue hill steppe, 4 VII 1959—Yun.); ***Jung. Alt.*** (Toli district, steppe, No. 4947, 4 VIII 1957—Kuan); ***Zaisan*** (Zimunai district, in Inkhotem river valley, on rather saline slopes, 1300 m, No. 10578, 25 VI 1959—Lee et al.).

General distribution: East. Sib. (south.), Far East (south.), North Mong., China (Dunbei), Korea (nor.)

47. C. **minutiscabra** Kük. ex V. Krecz. in Fl. SSSR, 3 (1935) 616, 380; B. Fedtsch. in Acta Horti Petrop. 38, 1 (1924) 212 (nom. nud.); Fl. Kazakhst. 2 (1958) 70.

Described from Kazakhstan. Type in Leningrad.

In spruce groves, about 3000 m above sea level.

IIA. Junggar: ***Tien Shan*** (valley of Muzart river, away from right peak of Tunu-daban near Oi-Terek area, spruce grove at high altitude covered with Pea shrub, 2850 m, 9 IX 1958—Yun.).

General distribution: Jung.-Tarb. (Jung. Alatau—Aktasty-tau hill).

Note. This species has been reported so far from only two sites. Related to C. *turkestanica* Rgl. but differing from it in more lax pistillate spikelets, sometimes with staminate flowers on top, and in very short beaks (0.5, not 1–1.2 mm) of utricles.

48. C. **relaxa** V. Krecz. in Fl. SSSR, 3 (1935) 616, 379; Egorova in Bot. mater. Gerb. Bot. inst. AN SSSR, 19 (1959) 81. —?C. *nitida* var. *filifolia* Kük. in Acta Hort. Gotoburg. 5 (1929) 44.

Described from East. Siberia (Transbaikal). Type in Leningrad. Plate III, fig. 1; map 11.

Mountain grass-herb and chee grass steppes and meadows, forest borders.

IA. Mongolia: ***Cent. Khalkha*** (Tumun-Degler somon, 7–8 km east of Erdeni-Gunyi khid, grass steppe, 29 VIII; Bain-Ula somon, 35–37 km south-east of somon, grass-tansy steppe, No. 16342, undated—1949—Yun.); ***East. Mong.*** (Manchuria railway station, steppe, 6 VI 1902—Litw.; Tamtsag-Bulak somon, 20 km south-west of Lag-Nur, chee grass steppe, 13 V 1944; 18–20 km nor.-west of Tamtsag-Bulak, chee grass steppe, 14 VIII 1949—Yun.; south of Tamtsag-Bulak, steppe, 18 VI 1956; 70 km east of Choibalsan, 14 VI 1954; east of Choibalsan, chee grass steppe, 24 VIII 1956; Khalkha-Gol somon, Khamar-Daban region, steppe, 18 VI 1954; 10 km west of Khabarg railway station, steppe, 1 VI 1955; 10 km south of Khabarg railway station, steppe, 20 V 1957; Matad somon, 15 km east of Shorvog, steppe, 5 VI 1956—Dashnyam).

IIIA. Qinghai: ***Amdo*** (Kadiger monastery, in pine forest, 25 V 1885—Pot.).

General distribution: East. Sib. (Dauria), North Mong. (Hang., Mong.-Daur.), China (North, North-west, South-west—nor. Sichuan).

49. C. **turkestanica** Rgl. in Acta Horti Petrop. 7, 2 (1880) 570; V. Krecz. in Fl. SSSR, 3 (1935) 377; Fl. Kirgiz. 2 (1950) 293; Fl. Kazakhst. 2 (1958) 70; Fl. Tadzh. 2 (1963) 126; p.p., excl. synon. C. *unguriensis* Litv. —C. *nitida* var. *aspera* (Boeck.) Kük. in Pflanzenreich [IV, 20], 38 (1909) 458; Krylov,

Fl. Zap. Sib. 3 (1929) 490. —C. *supina* auct. non Wahl.: Kük. l.c. 455, p.p., quoad pl. hymal. —C. *nitida* auct. non Host: Kük. l.c. 457, p.p. quoad pl. altaicae et Asiae med. et centr. —**Ic.:** Fl. Tadzh. 2, Plate XXVIII, figs. 7–10.

Described from Tien Shan (Kirgiz Alatau). Type in Leningrad. Plate III, fig. 3; map 9.

In grass-herb and grass steppes, on meadow slopes, in thin forests; from submontane plains to 3000 m above sea level but predominantly in middle mountain steppe belt.

IB. Kashgar: ***Nor.*** (21 km east of Aksu, low mountain, on slope, No. 1599, 15 VIII 1967—Kuan; west of Aksu, on Karachortszy hill, 3000 m, No. 8445, 19 IX; Muzart valley, 2750 m, No. 8274, 9 IX 1958—Lee and Chu; nor.-west. slope of Sogdantau mountain range, 2880 m, mountain-steppe belt, 19 IX 1958—Yun.); ***West.*** (valley of Sulu-sakhal river, 25 km east of Irkeshtam, on gentle slopes, 26 VII 1935—Olsuf'ev; in Koshkulak-Upal, 2800 m, No. 00243, 10 VI 1959—Lee et al.; Kingtau mountain range, nor. slope, 3–4 km south-east of Koshkulak settlement, upper part of forest belt, grassy spruce grove, 10 VI; same site, steppe belt, 10 VI 1959—Yun.).

IIA. Junggar: ***Altay region*** (from Karamai to Shara-Sume, No. 2645, 3 VIII; Shara-Sume, 1100 m, No. 2463, 26 VIII 1956—Ching; 20 km nor.-west of Shara-Sume, on Kran river, scrub-meadow steppe, 7 VII 1959—Yun.; Koktogoi, 1200 m, No. 1843, 13 VIII 1956—Ching; same site, 1470 m, No. 10386, 6 VI 1959—Lee et al.; in Koktogoi region, 1600 m, unnumbered, 17 VIII 1956—Ching): ***Jung. Alt.*** (Urtak-Sary, 19 VII 1878—Fet.); ***Tien Shan.***

General distribution: Jung.-Tarb., North and Centr. Tien Shan; Near East (?Afghanistan), Mid. Asia (West. Tien Shan, Pamiroalay), West. Sib. (Altay), Himalayas (west.—Kumaon).

Section Digitatae (Fries) Christ

50. **C. allivescens** V. Krecz. in Bot. mater. Gerb. Bot. inst. AN SSSR, 9 (1946) 190.

Described from North-west. China (Gansu province). Type in Leningrad.

In marshy forests.

IIIA. Qinghai: ***Amdo*** (Baga-gorgi river, in marshy forest, among moss, 2700 m, 26 V 1880—Przew.).

General distribution: China (North-west—Gansu province, single report).

Note. *C. allivescens* differs from all other species related to *C. pediformis* C.A. Mey. (section *Aunieria* V. Krecz.) in slender creeping rhizome. Nothing can be said about the utricles since they are almost totally undeveloped in the two available herbarium specimens of this species.

51. **C. aneurocarpa** V. Krecz. in Fl. SSSR, 3 (1935) 614, 369; Fl. Kirgiz. 2 (1950) 293; Fl. Kazakhst. 2 (1958) 73. —C. *pediformis* auct. non C. A. Mey.: Kük. in Pflanzenreich [IV, 20], 38 (1909) 490, p.p., quoad pl. Asiae med. et songor.

Described from Junggar (Jung. Alatau). Type in Leningrad. Plate IV, fig. 2; map 10.

On steppe and meadow slopes, predominantly southern, in subalpine steppes; in mountain-steppe belt, sometimes in forests (1000–2800 m above sea level).

IB. Kashgar: ***Nor.*** (nor.-west. slope of Sogdan-tau mountain range and valley of Taushkan-Darya river, hill-steppe belt, wormwood-cereal grass steppe, 2880 m, 19 IX 1958—Yun.); ***East.*** (4 km nor. of Turfan, on shaded slope, 2500 m, No. 5652, 15 VI 1958—Lee and Chu).

IIA. Junggar: ***Jung. Alt.*** (Khorgos, 1500–1800 m, 15 VI 1878—A. Reg., type!; Toli district, Bartok-Arba-Kezen', No. 1169, 6 VIII; same site, 1980 m, No. 1378, 8 VIII; Ven'tsyuan', on slope in forest, Nos. 2111 and 3415, 13 and 27 VIII 1957—Kuan; south-west. fringe of Maili mountain range, 40–42 km north-east of Kzyl-Tuz meteorological station toward Karaganda pass, hill-steppe belt, 14 VIII 1957—Yun. et al.); ***Tien Shan.***

General distribution: Balkh. Region (south.), Jung.-Tarb., North and Centr. Tien Shan; Mid. Asia (West. Tien Shan), West. Sib. (Kokchetau district).

Note. This species is very close to *C. pediformis* C. A. Mey. and possibly identical to it. V. Kreczetowicz distinguishes *C. aneurocarpa* from *C. pediformis* on the basis of more large veinless, 3.5–3.8 mm long utricles, staminate spikelet surpassing pistillate spikelets and obtuse glumes. A study of the entire material on *C. aneurocarpa* showed that its utricles are smaller in size than indicated by V. Kreczetowicz, i.e. (2.8) 3–3.3 (3.5) mm long. The utricles in the type specimen of *C. aneurocarpa* are 3–3.1 mm long. The absence of veins in the utricles is a consistent feature. Plants with veinless utricles are found, however, even among *C. pediformis* in different parts of its range. The latter species in general greatly varies in the extent of venation of utricles. It is possible that there is some pattern in the distribution of *C. pediformis* with veinless utricles. Establishing this aspect calls for a thorough study of extensive material on this species, covering its entire distribution range, which essentially falls outside the confines of Central Asia.

A staminate spikelet standing above the level of the pistillate spikelet, albeit a distinctive feature for *C. aneurocarpa*, is also equally characteristic of *C. pediformis*. A staminate spikelet at the same level as the upper pistillate spikelet or slightly below it is also seen in these two species. If, however, the uppermost pistillate spikelet is not developed, a rather frequent character, the staminate spikelet will obviously surpass the pistillate spikelet next to it. This feature can be found among plants standing on the same mat. Depending on the presence or absence of the upper pistillate spikelet, the peduncle of the staminate spikelet may be shorter or longer.

Glumes in *C. aneurocarpa* as well as in *C. pediformis* vary in shape of the tip from obtuse to rather acute and acuminate.

Neither species can be distinguished by length of the peduncle of pistillate spikelets nor beak of the utricles.

Thus, there is only one distinctive feature of *C. aneurocarpa*, i.e. absence of venation in the utricles, although utricles without veins are encountered among *C. pediformis* too. Among Central Asian specimens of the latter, veins on the utricles are manifest to a large extent only in the rear part and the utricles themselves are slightly larger than in *C. aneurocarpa* and have a more extended base.

52. **C. lanceolata** Boott var. **alaschanica** Egor. var. nov. A var. typica differt: rostris utriculorum longioribus, squamis spicularum pistillatarum ovatis utriculis subaequilongis, necnon caulibus paulo scabris.

Typus: China, jugum Alaschan, in angustis Tszosto, in declivibus septentrionalibus, No. 146, 11 V 1908, S. S. Czetyrkin. In Herb. Inst. Bot. Acad. Sci. URSS (Leningrad) conservatur.

Plate IV, fig. 1.

Along slopes and in lower and middle mountain zones.

IA. Mongolia: ***Alash. Gobi*** (Alashan mountain range and gorge, Tsuburgan-Gol, 29 IV; same site, Yamata gorge, nor. and east. slopes, middle and lower belt, 6 V; same site, Tszosto, nor. slope, on humus soil, No. 146, 11 V 1908, Czet., type!).

General distribution: China (North).

Note. Plants treated by us as var. *alaschanica* m. were marked for description as a species in 1938 by V. Kreczetowicz. The specimen selected by Kreczetowicz as the type served as the type for var. *alaschanica*. From among the above-cited specimens from Alashan mountain range, only the type specimen has relatively mature utricles. The utricles of the other specimens are hardly developed. It cannot, therefore, be said whether they possess the basic distinctive feature of the variant, viz. a very long beak.

It is possible that the plants cited here as var. *alaschanica* could be regarded as an independent species. A large amount of material needed, however to establish this taxon.

N.A. Ivanova identified Alashan plants as C. *lanceolata* var. *subpediformis* Kük., which is hardly correct. This variant is found, according to Kükenthal, in Central China and Japan. Kükenthal describes it [Pflanzenreich [IV, 20], 38 (1909) 493] as possessing, compared to the type form C. *lanceolata* Boott, very dense pistillate spikelets and obovate glumes slightly exceeding the utricles. We did not see authentic specimens of var. *subpediformis* but consider it a form with very short and wider glumes not infrequently found in C. *lanceolata*. Further, plants with such glumes exhibit no other differences from plants with elongated glumes. Several intermediate forms exist between the extreme types of glumes in C. *lanceolata*.

It must be pointed out that V. Kreczetowicz analysed some specimens from Northern and North-western China. This material turned out to be highly heterogeneous. Plants from North China (Bunge's collection) have beaks that are average in size between var. *alaschanica* and the typical form of C. *lanceolata*. They stand closer to var. *alaschanica* in the extent of rugosity of stem and form of glumes. Another specimen from Northern China (G.N. Potanin's collection), in our view, belongs to C. *pediformis*.

Finally, plants from North-western China (Gansu province, G.N. Potanin's collection) belong, in all probability, to C. *laeta* Boott var. *rostrata* Kük. They differ sharply from C. *lanceolata* in elongated, straight, shrot-biscuspid beaks and also low pistillate spikelets with long stalks right at the base of the shoots.

53. **C. macroura** Meinsh. in Acta Horti. Petrop. 18 (1901) 404; V. Krecz. in Fl. SSSR, 3 (1935) 367; Kitag. Lin. Fl. Mansh. (1939) 106; Grubov, Consp. fl. MNR (1955) 86; Fl. Kazakhst. 2 (1958) 72. —C. *pediformis* var. *macroura* (Meinsh.) Kük. in Oefvers. Finska Vet. Soc. Förh. 45, 8 (1902–1903) 10; idem, Pflanzenreich [IV, 20], 38 (1909) 491; Krylov, Fl. Zap. Sib. 3 (1929) 494. —**Ic.:** Fl. SSSR 3, Plate XIX, fig. 4.

Described from West. Siberia. Type in Leningrad.

In spruce-larch forests.

IIA. Junggar: ***Altay Region*** (Qinhe [Chingil'], on slope, 1900 m, No. 833, 2 VIII; Altay, 1400 m, No. 2524, 27 VIII 1956—Ching; 30 km nor. of Koktogoi, right bank of Kairta river, valley of Kuidyn river, mixed spruce-larch forest, 15 VII 1959—Yun.; Altay hills in Chintas region, in forest, 1800 m, No. 10729, 21 VII 1959—Lee et al.).

General distribution: Jung.-Tarb.; Urals, West. Sib. (south), East. Sib., Far East (south., very rare), North Mong. (Hent., Hang.), China (Altay, Dunbei).

54. **C. pediformis** C. A. Mey. in Mém. Ac. Sci. St.-Pétersb. Sav Étr. 1 (1831) 219; Forbes and Hemsley, Index Fl. Sin. 3 (1905) 303, p.p.; Kük. in Pflanzenreich [IV, 20], 38 (1909) 490, p.p.; V. Krecz. in Fl. SSSR, 3 (1935) 368; Grubov, Consp. fl. MNR (1955) 87; Fl. Kazakhst. 2 (1958) 72. —**Ic.:** C. A. Mey. l.c., Tab. X; Fl. SSSR, 3, Plate XIX, fig. 3.

Described from East. Siberia. Type in Leningrad.

On steppe and meadow slopes, in steppes, on forest fringes, in sparse forests and scrub; up to alpine belt.

IA. Mongolia: ***Khobd., Mong. Alt.*** (west. slope of Indertiin-Nuru in upper Tsagan-Gol river [right bank of Buluguna tributary], alpine steppe, 23 VII; south. flank of Indertiin-Gol river valley, hill steppe, 24 VII; ascent on Burgastyn-daba pass in upper Indertiin-Gol, hill steppe, 6 VIII 1947—Yun.; nor. slope of Khan-Taishir mountain range 15 km south-east of Yusun-Bulak, sedge-fescue meadow along forest border, 1 IX 1948—Grub.); ***Cis-Hing.*** (Yaksha railway station, on rocks and hill slopes, Nos. 2198, 2253 (1954)—Wang; Khalkha-Gol somon, Salkhit-ula in Dege-Gola region, tansy steppe, 9 VI; same site, Numurgin-Gol river, opposite Khabargi mountain, herb-wheat grass floodplain meadow, 11 VI 1956—Dashnyam); ***Cent. Khalkha*** (180 km south-south-west of Ulan-Bator town, Sorgol Khairkhan hill, along cleavages in granite, 16 V 1941 and 15 VIII 1943; Buyantu somon, midcourse of Ara-Dzhirgalante-Gol, fescue-herb mountain steppe, 2 VII 1949—Yun.); ***East. Mong.*** (Manchuria railway station, steppe, 6 VI 1902—Litw.; same site, 1915—Nechaeva; same site, Nanshan hill, 900 m, 24 VII 1951—Wang; Kharkhonte railway station, sands, 7 VI 1902—Litw.; Dzodol-Khan somon, nor.-west of extinct volcano at Dzodol-Khan, on basalt faces, 15 VI 1944—Yun.; Khuntu somon, 5 km west of Toge-Gol lake, herb meadow, 7 VIII 1949—Yun.; Khalkha-Gol somon, flood-plain of Mogoityn-Gol river, herb-sedge meadow, 12 VI; same site Khairkhant hill top, among rocks and stones, 14 VI; Ara-Dzhargalant somon, Tsog-Delger hill top, among rocks and stones, 26 VI 1956—Dashnyam); ***Gobi-Alt.*** (Baga Bogdo, grassy slopes between canyons, 1800–2100 m, No. 216, 1925—Chaney; Dzun-Saikhan hills, among willow thickets on slope, 26 VIII 1931—Ik.-Gal.; same site, steppe slope, 19 VI 1945—Yun.; Ikhe-Bogdo mountain range, nor. slope, Tsagan-Burgas creek, subalpine steppe, 15 IX 1943—Yun.); ***Alash. Gobi*** (Alashan mountain range, Yamate gorge, nor. and east. slopes in middle and lower zones, No. 101, 1907–1909—Czet.).

IIA. Jungg: ***Jung.-Gobi.***

General distribution: Arct. (Europ., Asiat., very rare), Europe (Europ. USSR—Cent. Russian upland and east. regions, Urals), West. Sib. (south.), East. Sib., Far East (Amur Region), North Mong., China (Dunbei, Nor.).

55. **C. supermascula** V. Krecz. in Bot. mater. Gerb. Bot. inst. AN SSSR, 9 (1946) 188; grubov, Consp. fl. UNR (1955) 88. — *C. pediformis* var. β. *macroura* (Meinsh.) Kük. in Pflanzenreich [IV, 20], 38 (1909) 491, p.p., quoad pl. amur. —*C. sutschanensis* auct. non Kom.: V. Krecz. in Fl. SSSR, 3 (1935) 368. —*C. macroura* auct. non Meinsh.: Egorova in Bot. mater. Gerb. Bot. inst. AN SSSR, 19 (1959) 81.

Described from the Far East (Amur basin). Type in Leningrad.

On rocky slopes, in sparse forests.

IA. Mongolia: ***Cis-Hing.*** (B. Hinggan foothills, left bank of Numurgin-Gol river, steep rocky slopes, 5 V 1944—Yun.: Arshan town, dry site on hill meadow, No. 316, 14 VI 1950—Chang); ***East. Mong.*** (near Khailar railway station, forest on sands, 10 VI 1902—Litw.).

General distribution: East. Sib. (Transbaikal), Far East (south.), Nor. Mong. (Hent.), China (Dunbei, South-west.—Sichuan province), Korea.

Note. *C. supermascula* is closely related to the Siberian species *C. macroura* Meinsh. It is well distinguished from the latter in very dense beds. The sheath of the lower bract leaf (i.e., bract leaf of the second pistillate spikelet from the top) has a relatively developed blade 1–3 cm long, which varies from setaceous to narrowly linear. The sheath or lower bract leaf in *C. macroura* has a short-acuminate, scaly blade. *C. macroura* inhabits predominantly forests, less often open expanses such as forest fringes and dry slopes. *C. supermascula* grows in dry meadows, steppe and rocky slopes and in thin forests. Both the species have extensive distribution ranges.

Section Paniceae Tuckerm. ex Kük

56. **C. panicea** L. Sp. pl. (1753) 977; ?Forbes and Hemsley, Index Fl. Sin. 3 (1905) 302; Kük. in Pflanzenreich [IV, 20], 38 (1909) 510; Krylov, Fl. Zap. Sib. 3 (1929) 505; V. Krecz. in Fl. SSSR, 3 (1935) 344; Fl. Kirgiz, 2 (1950) 292; Fl. Kazakhst. 2 (1958) 71; Fl. Tadzh. 2 (1963) 125. —**Ic.:** Fl. SSSR, 3 Plate XX, fig. 5.

Described from Sweden Type in London (Linn.).

In wet and marshy areas.

IIA. Junggar: ***Tien Shan*** (between Saryk and Mengute, 1800–2100 m, 27 VII 1879—A. Reg.; near Kul'dzha [IV-VII 1877], A. Reg.).

General distribution: Balkh. Region, Jung.-Tarb., Nor. and Cent. Tien Shan; Europe, Asia Minor, Caucasus, Mid. Asia (Pamiroalay-Zeravshan mountain range, ?West. Tien Shan), West. Sib. (south.), East. Sib. (south., very rare), ?China (North), North Amer. (introduced), North Africa.

Section Frigidae Fries ex Kük.

57. **C. alajica** Litw. in Tr. Bot. muzeya AN, 7 (1910) 99; V. Krecz. in Fl. SSSR, 3 (1935) 355; Fl. Kirgiz. 2 (1955) 292; Fl. Tadzh. 2 (1963) 125. —**Ic.:** Fl. Tadzh. 2, Plate XXVII, fig. 4.

Described from Mid. Asia (Alay valley). Type in Leningrad. Plate V, fig. 1; map 10.

On meadow and steppe slopes in subalpine belt and in upper forest limit.

IB. Kashgar: ***Nor.*** (valley of Muzart river, first downward turn Lyanger 5–7 km from Kurgan settlement, subalpine belt, meadow, slopes with spruce grove, 12 IX 1958—Yun.); ***West.*** (on Kashgar to Tashkurgan road, in Koshkulak-Upal, on nor. slope, No. 00225, 10 VI 1959—Lee et al.; Kingtau mountain range, 3–4 km south-east of Koshkulak settlement, steppe belt, nor.-west. slope, among juniper groves, 10 VI 1959—Yun.).

General distribution: East. Pam. (Kainda valley); Mid. Asia (Pamiroalay—Turkestan mountain range, Transalay mountain range, Alay valley).

Note. This species was earlier known only from the USSR territory. The original diagnosis of *C. alajica* Litw. contained an error—the uppermost spikelet was described as staminate. Actually, the uppermost spikelet in this species as well as the upper lateral spikelets closely adjoining the uppermost one are androgynous and only the lower spikelet is usually wholly pistilate. This error, repeated by V. Kreczetowicz in "Flora of the USSR," led to incorrect representation of the affinities of *C. alajica*. V. Kreczetowicz placed the latter in section *Campylorhina* V. Krecz. among the constituents of which—*C. vaginata* Tausch, *C. michelii* Host, *C. longirostrata* C. A. Mey. and others—androgynous spikelets are never found. *C. alajica* should be placed in section *Frigidae* Fries ex Kük., subsection *Decorae* Kük. [section *Orphinascus* (Boern.) V. Krecz., cycle *Palaeorphinascus* V. Krecz.]. Most species of this subsection are distributed in the Himalayas. In the flora of the USSR, this group is represented by *C. alajica* and Caucasian species *C. pontica* Albov.

58. **C. alexeenkoana** Litv. in Tr. Bot. muzeya AN, 7 (1910) 98; V. Krecz. in Fl. SSSR, 3 (1935) 338; Persson in Bot. Notis. (1938) 276; Fl. Kirgiz. 2 (1950) 292; Fl. Kazakhst. 2 (1958) 68; Fl. Tadzh. 2 (1963) 124. —*C. macro-*

gyna auct. non Steud.: Polyak. in Fl. Kazakhst. 2 (1958) 65. —**Ic.:** Fl. Tadzh. 2, Plate XXVII.

Described from Pamir (Tagdumbash-Pamir). Type in Leningrad. Plate VI, fig. 3; map 11.

Along rock and rubble slopes of steppes, rock fissures, short-grass meadows along brooks, carex-kobresia meadows, juniper groves; in high-altitude belt and at upper forest boundary.

IB. Kashgar: *Nor.* (Uchturfan, 4 and 18 IV, 23 V 1908—Divn.; Chzhaosu district, Aksu region, on slope, No. 3499, 14 VIII 1957—Ching; south. slope of Tien Shan in Kuchi region, 2620 m, Nos. 10033 and 9971, 26 VII 1959—Lee et al.); *West.* (King-tau mountain range, nor. slope, 4 km south-east of Koshkulak settlement, 10 VI 1959—Yun.; same site, spruce grove with juniper-kobresia, 10 VI; south. Tien Shan, 3–4 km south-west of Torugart settlement, on USSR border, high-altitude steppe, 3600 m, 20 VI 1959— Yun.; between Kashgar and Torugart, on hill top, No. 09747, 20 VI 1959—Lee et al.); *South.* (Cherchen district 10 km south-east of Achan, on nor. slope, 3460 m, No. 9399, 2 VI; same site, on slope of Achan hill, No. 9379, 1 VI; 30 km south-east of Achan, high-altitude belt, 4210 m, No. 9418, 4 VI 1959—Lee et al.); *East.* (in Turfan, on sheded slope, 2500 m Nos. 5787 and 5810, 23 VI; Tsakhanur pass in Khotun-Sumbule, on valley alluvium, 3100 m, No. 6512, 15 VIII 1958—Lee et al.).

IIA. Junggar: *Tien Shan* (very common).

IIIB. Tibet: *Chang Tang* (Keriya mountain range, Kyuk-Egil' river, in alpine meadow, 10 and 11 VII 1885—Przew.).

IIIC. Pamir.

General distribution: Jung.-Tarb. (Jung. Alatau), Nor. and Cent. Tien Shan, East. Pam.; Mid. Asia (West. Tien Shan and Pamiroalay).

Note. This Central Asian plant is very close to the Siberian C. *macrogyna* Turcz. ex Steud. V. Kreczetowicz (l.c.) erroneously placed these species in different sections. The main difference between these species is that the peduncles of pistillate spikelets in C. *alexeenkoana* Litw. do not emerge from sheaths of bract leaves (spikelets appear sessile) while, in C. *macrogyna*, the peduncles usually emerge 1–2.5 (3) cm. Moreover, the utricles in the former species are usually oblong-obovate or obovate and cuneate at the base while, in the latter, they are oblong-ovate and ovate and orbicular at the base. The two species grow under similar ecological conditions.

59. **C. atrofusca** Schkuhr, Riedgr. 1 (1801) 106; Kük. in Pflanzenreich [IV, 20], 38 (1909) 553; p.p., Ostenf. in Hedin, S. Tibet (1922) 92; Krylov, Fl. Zap. Sib. 3 (1929) 506; Pampanini, Fl. Carac. (1930) 85; V. Krecz. in Fl. SSSR, 3 (1935) 282; Hao in Englers Bot. Jahrb. 68 (1938) 584. —*C. ustulata* Wahl. in Svensk. Vet. Ac. Handl. 24 (1803) 156; Hemsley, Fl. Tibet (1902) 202; Forbes and Hemsley, Index Fl. Sin. 3 (1905) 316. —*C. stilbophaea* V. Krecz. l.c. 605, 283; Grubov, Consp. fl. MNR (1955) 88. —*C. oxyleuca* V. Krecz. l.c. 605, 284; Persson in Bot. Notis. (1938) 276; Fl. Kirgiz. 2 (1950) 287; Fl. Kazakhst. 2 (1958) 64; Fl. Tadzh. 2 (1963) 116. —**Ic.:** Fl. Tadzh. 2, Plate XXV, fig. 1 (as C. *oxyleuca* V. Krecz.).

Described from Europe. Type in Galle (?). Plate VI, fig. 4.

In wet and marshy grass plots, Carex-Kobresia meadows, along river banks; in high-altitude belt.

IA. Mongolia: *Mong. Alt.*

IB. Kashgar: *Nor.* (Uchturfan, 18 VI 1908—Divn.); *South.* (Keriya, Chivei area, alpine meadow, 3150–3900 m, 15 VII 1885—Przew.).

IC. Qaidam: ***Mount.*** (Sarlyn hills [Kurlyka region], Kuku-Bulak area, around a spring, 30 V 1895—Rob.).

IIA. Junggar: ***Tien Shan*** (Muzart pass, 3000–3450 m, 18 VIII 1877; Aryslyn, 2700–3000 m, 11 and 16 VII 1879; between Sumbe and Kazan, 12 VII 1879; Alysi spring [Alysi Quelle], 2700–3000 m, 11 IX 1880—A. Reg.; upper Ulan-Usu river, 8–10 km above Dzhartas falls on road to Danu pass, high-altitude zone, marshy meadow, 19 VII; same site, valley of Danu-Gol river 7–8 km below Se-Daban pass, high-altitude zone, Carex-Kobresia meadow, 21 VII; same site, upper Danu-Gol river, marshy Carex meadow, 22 VII 1957—Yun.; M. Yuldus, 2–3 km west of Kotyl' pass, on road from Yuldus to Karashar, high-altitude zone, Carex marshy meadow, 15 VIII 1958—Yun.; Danu-Gol river, on slope, 2500 m, Nos. 2066 and 2084, 21 VII; 1.5–3 km south of Danu, in a gorge, on slope, 3150 m, Nos. 344 and 414, 21 VII; 7–8 km south of Danu, 3450 m, No. 520, 22 VII; Danu-Gol river, hill meadow, 2200 m, No. 2214, 23 VII 1957—Kuan; from Bort to eastern Taskhan canal in Khomote, on slope, 2780 m, No. 7083, 5 VIII; Taskhan canal, in marsh, 3280 m, No. 7105, 6 VIII; Malyi Luitsichen in Khomote, on south. slope, 3500 m, No. 7163, 9 VIII 1958—Lee and Chu; "Thian-Shan, Köl, about 2700 m, 2 VIII 1932"—Persson, l.c.).

IIIA. Qinghai: ***Nanshan*** (Rakogol river, on alpine meadow, [21 and 22 VII], 1880—Przew.; Sharagol'dzhin, Baga-Buran mountain range, 15 VI; Nanshan mountain range, alpine belt, 3450 m, 11 VII 1894—Rob.); ***Amdo*** ("am Ufer des Sees Gan-gia-zo, 3440 m, No. 1001, 24 VIII; Tsi-gi-gan-ba, auf feuchten Alpenwiesen des Gebirges Ming-ke-schan, 4500 m, No. 1034, 27 VIII 1930"—Hao, l.c.).

IIIB. Tibet: ***Chang Tang*** ("Nameless valley in the interior of E. Tibet, 33°32′ N, 88°52′ E, 5127 m, 18 VIII 1901, S. Hedin"—Ostenfeld, l.c.); ***Weitzan*** (Yangtze river, 3900 m, 21 VI 1884—Przew.; Burkhan-Budda mountain range, nor. slope, Nomokhun gorge, 4200 m, 22 V 1900—Lad.; valley of Alyk-Norin-Gol river, 3630 m, 7 VI 1901—Lad.; "Amne Matchin, in Alpenwiesen, No. 1101, 2 IX 1930"—Hao, l.c.); ***South.*** (Goring valley, 30°12′ N, 90°25′ E, 4800 m, Littledale"—Hemsley, l.c.; "Balch pass, 5100 m, Strachey and Winter-bottom"—Hemsley, l.c.).

IIIC. Pamir (Kenkol river, Tom-Kara area, marshy bank of brook, 14 VI; Ulugtuz gorge in Charlysh river basin, along wet meadow, 28 VI 1909—Divn.; Kokat pass region, 3800–4000 m, 16 VI Pil'nen gorge, 4500–5000 m, 1 VII; Taspetlyk area, 4000–5000 m, 25 VII; Goodzhiro river, 4500–5500 m, 27 VII; Shorluk gorge, 4000–5500 m, 28 VII 1942—Serp.).

General distribution: Jung. Tarb., Nor. and Cent. Tien Shan, East. Pam.; Arct. Europe (hills of Cent. Europe, Scotland and Scandinavia, Urals), Near East (Afghanistan), Mid. Asia (West. Tien Shan), West. Sib. (Altay), East. Sib., Far East (sea of Okhotsk coast), North Mong. (Fore Hubs., Hang.), China (North-west., South-west.), Himalayas, North Amer.

Note. *C. oxyleuca* V. Krecz. identified by us as *C. atrofusca* Schkuhr has been described from Fergana (Ters-Agar pass). According to V. Kreczetowicz, the distribution range of *C. oxyleuca* includes Tien Shan, Junggar Alatau, Tarbagatai, Pamiroalay, Kashgar, Afghanistan and Kashmir. A comparison of specimens identified by V. Kreczetowicz as *C. oxyleuca* with *C. atrofusca* revealed no differences between these species. The white membranous tips of glumes of pistillate spikelets, by which the original author mainly distinguished his species, are found (and not infrequently) in *C. atrofusca* throughout its range and are developed to varying degrees.

Further, contrary to V. Kreczetowicz's reference (l.c.), the peduncles of pistillate spikelets in *C. oxyleuca* do not differ in thickness from the corresponding feature in *C. atrofusca*, while the spikelets themselves are less drooping and denser than in the latter. Spikelets of *C. atrofusca* droop differently, and quite often they are suberect; this species also varies in the extent of spacing of spikelets. There is, therefore, no justification to regard *C. oxyleuca* as an independent species.

V. Kreczetowicz himself later identified *C. stilbophaea* V. Krecz., described from the Sayan range, as *C. atrofusca*, as pointed out in his much later identification of the corresponding herbarium specimens.

Thus, the distribution range of arctic-alpine C. *atrofusca* is considerably wider than indicated in "Flora of the USSR".

V. Kreczetowicz isolated the herbarium specimens from Qaidam, Qinghai, Tibet, Nor.-west. and South-west. China into a distinct new species. Some of these specimens have very large pistillate spikelets and lanceolate glumes with light-coloured veins and sometimes very broad leaves. These characteristics have not been correlated. In our view, plants possessing these characteristics should be regarded as a form of C. *atrofusca*.

60. **C. coriophora** Fisch. et Mey. ex Kunth, Enum. pl. 2 (1837) 463; V. Krecz. in Fl. SSSR, 3 (1935) 282; Kitag. Lin. Fl. Mansh. (1939) 99; Grubov, Consp. fl. MNR (1955) 83. —C. *ustulata* var. *pallida* Meinsh. in Acta Horti. Petrop. 18 (1901) 357; Danguy in Bull. Mus. nat hist. natur. 20 (1914) 145. —C. *atrofusca* var. *coriophora* (Kunth) Kük. in Pflanzenreich [IV, 20], 38 (1909) 554. —C. *atrofusca* var. *fusca* Kryl. Fl. Alt. 6 (1912) 1514; idem, Fl. Zap. Sib. 3 (1929) 507. —**Ic.:** Fl. SSSR, 3, Plate XVII, fig. 3.

Described from East. Siberia (Transbaikal). Type in Leningrad.

In wet and marshy meadows within the forest belt.

IA. Mongolia: ***Cis-Hing.*** (Khuntu somon, 4–5 km west of Khalzan-ul, ravine floor, herb meadow, 8 VIII 1949—Yun.; "King-hans, vallée du Djatangol, bords de la riviére, 28 VI 1896, No. 1401, Chaff."—Danguy, l.c.); ***Cent. Khalkha, East. Mong.***

IIA. Junggar. Tien Shan (Dzhagastai-Gol, 2700 m, 5 IX 1879—A. Reg.).

General distribution: West. Sib. (Altay), East. Sib., North Mong., China (Dunbei).

61. **C. dielsiana** Kük. in Notes Bot. Gard. Edinburgh, 8 (1913) 10; Walker in Contribs. U.S. Nat. Herb. 28 (1941) 599.

Described from South-west. China (Yunnan province). Type in Edinburgh (?).

IIIA. Qinghai: ***Nanshan*** ("La Ch'iung Kou [Hsi Ning Hsien], in large tufts on exposed, moist, grassy slopes or in woods, common, No. 613, 1923, R.C. Ching"—Walker, l.c.).

General distribution: China (South-west.).

62. **C. griffithii** Boott in Trans. Linn. Soc. 22, 3 (1851) 138; Litv. in Tr. Bot. muzeya AN, 7 (1910) 91; V. Krecz. in Fl. SSSR, 3 (1935) 281; Fl. Kirgiz. 2 (1950) 286; Fl. Kazakhst. 2 (1958) 64; Fl. Tadzh. 2 (1963) 113. —C. *nivalis* auct. non Boott: Kük. in Pflanzenreich [IV, 20], 38 (1909) 551, p.p., quoad pl. Asiae med. —C. *nivalis* f. *griffithii* (Boott) Kük. l.c. 553. —C. *nivalis* var. *griffithii* (Boott) T. Koyama in Acta Phytotax. and Gebot. 17, 4 (1958) 100. —**Ic.:** Fl. SSSR, 3, Plate XVII, fig. 2.

Described from the Near East (Afghanistan). Type in London (K). In high-altitude belt.

IB. Kashgar: West. (Kashgar, below Dzhitym pass, 28 VII 1903—Lipsky).

General distribution: Nor. and Cent. Tien Shan, East. Pam.; Near East, Mid. Asia (Pamiroalay, West. Tien Shan), Himalayas (west.).

63. **C. przewalskii** Egor. sp. nov. —?C. *haematostoma* var. *submacrogyna* Kük. in Pflanzenreich [IV, 20], 38 (1909) 561.—Planta perennis, 15–40 cm alt., laxe caespitosa, rhizomate repenti. Caules ca. 1 mm in diam., sulcati, leves vel paulo scabri. Folia caulem subaequantia, 2–3 mm lt., plana. Inflorescentia 3–5 cm lg., e spiculis 3–7 approximatis constans. Bractea in-

feriora cum vagina 0.7–1.5 cm lg. et lamina angustiore inflorescentiam subaequantia. Spiculae superiores (1–3) vel staminatae omnino vel basi cum floribus pistillatis, 1–1.5 cm lg. Spiculae pistillatae in numero 2–4, 1–2 cm lg., 5–7 mm lt., superiores subsessiles, inferiores pedunculatae, pedunculis 1–1.5 cm lg.; spiculae inferiores interdum ramosae. Squamae spicularum pistillatarum oblongo-ovatae, acutae, castaneae, utriculorum brevius. Utriculi compresso plano-convexi, elongato-ovati, 6 mm lg., 1.8 mm lt., membranacei, flavo-albi, superne purpureo-brunnei, antice nervis 5–6 tenuibus, postice enervis, superne marginibus paulo scabri, in rostrum antice plus minusve emarginatum apice albo-membranaceum sensim angustati. Stigmata 2. Fructus lateovatus, 1.5 mm lg., 1.2 mm lt.

Typus: China, Tibet, prov. Tsinghai, 108 km ad occidentem a Sinin, et 6 km ad occidentem pag. Daudanche, in steppa montana, 3400 m s.m., 5 VIII 1959, M.P. Petrov. In Herb. Inst. Bot. Acad. Sci. URSS (Leningrad) conservatur.

Affinitas. A specie plus minusve proxima C. *haematostoma* Nees stigmatibus 2 (nec 3), utriculis plano-convexis et superne marginibus baulo scabri (nec trigoni et marginibus hispidi), squamis acutis (nec breviter aristatis), necnon rostris apice emarginatis 0.25 mm lt. (non bidentatis, 0.4–960.5 mm lt.) bene differt.

Plate V, fig. 2; map 11.

IIIA. Qinghai: ***Nanshan*** (Tetung river, in alpine zone, 28 VIII 1872—Przew.; 70 km southeast of Chzhan'e town, in hill valley, on grass plot with shrubs, 2600 m, 12 VII 1958; 108 km west of Sinin town, 6 km west of Daudanhe settlement, scrub hill steppe, 3400 m, 5 VIII 1959—Petr. type!).

64. **C. macrogyna** Turcz. ex Steud. Syn. Cyper. (1855) 236; Bess. in Flora, 17, 1, Beibl. (1834) 27 (nom. nud.); Turcz. in Bull. Soc. natur. Moscou, 11, 1 (1838) 104 (nom. nud.); Kük. in Pflanzenreich [IV, 20], 38 (1909) 560; Krylov, Fl. Zap. Sib. 3 (1929) 508; V. Krecz. in Fl. SSSR, 3 (1935) 293; Fl. Kirgiz. 2 (1950) 288; Grubov, Consp. fl. MNR (1955) 85. —**Ic.**: Kük. in Russk. bot. zh. 3–6 (1911), fig. 125.

Described from East. Siberia (South. Transbaikal). Type in Paris. Isotype in Leningrad. Plate VI, fig. 2.

Mountain steppes, stone slopes, Carex-Kobresia meadows, short-grass meadows along banks of brooks, spruce and juniper groves; from upper forest belt to high-altitude zone.

IA. Mongolia: ***Gobi-Alt.*** (Baga Bogdo, on grassy slopes, 1800–2100 m, No. 234, 1925—Chaney; Ikhe-Bogdo mountain range, Burgas upper creek, high-altitude belt, carex-kobresia meadow, 6 IX; same site, top of Ikhe-Bogdo mountain range, 11 IX 1943—Yun.; same site, south. slope of Ikhe-Bogdo mountain range, Narin-Khurimt creek, fescue-sedge upland steppe, 28 VI; same site, plateau in Ikhe-Khurimt upper creek, fescue-kobresia high-altitude steppe on placers, 28 VI; same site, Shishkhid upper creek, high-altitude belt, dry Kobresia meadow, 30 VI; same site, Sevs-ul upper creek, stone-covered steppe, lower hill belt, 2 VII; same site, Tsagan-Gol river near somon, along bank, 3 VII 1945—Yun.; same site, Narin-Khurimt gorge, on rocks, 2900 m, 28 VII; Narin-Khurimt upper gorge, site among rocks, south. slope, 29 VII 1948—Grub.).

IIA. Junggar: ***Tien Shan*** (Naryn-Gol, Tsagan-Usu tributary, 10 VI 1879—A. Reg.).

General distribution: Arct. (Asiat.), West. Sib. (Altay-Narym mountain range), East. Sib., Far East (Okhotsk coast), North Mong. (Force Hubs., Hang), China (North).

65. **C. montis-everesti** Kük. in Kew Bull. Misc. Inform. 6 (1934) 261.

Described from Tibet. Type in London (K).

IIIB. Tibet: ***South.*** ("Tibet, Mt. Everest, Camp I [Rongbuck Valley], in moraine, 5460 m, 2 VII 1933, R.L. Wager, type!"—Kük. l.c.).

Note. This species is known so far only from the cited location. The sedge regarded as *C. montis-everesti* in "Flora of Tadzhikistan" (2, 1963) belongs to another species.

66. **C. nivalis** Boott in Trans. Linn. Soc. 20, 3 (1851) 136; Duthie in Alcock, Rep. Pamir Commiss. (1898) 27; Strachey, Catal. (1906) 202; Kük. in Pflanzenreich [IV, 20], 38 (1909) 551, p.p. (excl. pl. Asiae med. et synon. *C. griffithii* Boott); Ostenf. in Hedin, S. Tibet (1922) 92; Pampanini, Fl. Carac. (1930) 84. —*C. lepusaestatis* T. Koyama in Acta Phytotax. and Geobot. 16 (1957) 167. —**Ic.:** Pflanzenreich [IV 20], 38, fig. 91.

Described from the Himalayas. Type in London (K).

In alpine meadows.

IIIC. Pamir ("Pamir region, 3900–4200 m, No. 17773, Alcock" Duthie, l.c.; "Eastern Pamir, Mus-tagh-ata, Jambulak-bashi, 4439 m, 16 VIII 1894, S. Hedin"—Ostenf. l.c.).

General distribution: Near East (Afghanistan, Pakistan), Himalayas (west., Kashmir).

67. **C. stenocarpa** Turcz. ex V. Krecz. in Fl. SSSR, 3 (1935) 607, 291; Fl. Kirgiz. 2 (1950) 287; Grubov, Consp. fl. MNR (1955) 88; Fl. Kazakhst. 2 (1958) 65; Fl. Tadzh. 2 (1963) 116. —*C. sempervirens* ssp. *tristis* (M.B.) Kük. in Pflanzenreich [IV, 20], 38 (1909) 569, p.p. —*C. tristis* auct. non M.B.: Kryl. Fl. Zap. Sib. 3 (1929) 509; Pampanini, Fl. Carac. (1930) 85.

Described from East. Siberia (South. Transbaikal). Type in Leningrad. Plate VI, fig. 1; map 12.

In subalpine and alpine meadows, and larch and juniper groves in upper forest limit.

IA. Mongolia: ***Khobd.*** (Bugusun-daba, Sailyugem, 24 VI 1869—Mal'chevskii; Kharkhira river, on pebble bed, 24 VII; Tszusylan river, in forest 13 VII 1879—Pot.; Tolbo-Kungei Alatau mountain range, 3200 m, high-altitude belt, kobresia meadow, 5 VIII 1945—Yun.); ***Mong. Alt.*** (Taishiri Ola, 16 VII 1894; slopes of Shadzagain-Suburga pass, 22 VII; slopes of upper Naryn river, larch forest, 23 VII 1898—Klem.; 8 km from Dzasaktu-Khana camp, larch forest on nor. slope, 9 VIII 1930—Pob.; Khan-Taishiri, nor. slope near peak, 15 VIII and 21 IX 1945—Leont'ev; Bulugun somon, Kharagantin-daba, alpine meadow, 23 VII; same site, 2 km west of somon on road to Kharagaitu-khutul, alpine steppe, 24 VII; same site, south. slope, Indertiin-Gol—Tumurtu-Gol pass, alpine meadow, 25 VII; Bulugun river basin, upper Ketsu river, alpine meadow, 26 VII 1947—Yun.); ***Gobi-Alt.*** (Ikhe-Bogdo mountain range, Narin-Khurimt-ama and Ketsu-ama water divide, kobresia-carex meadow on rock debris, 28 and 29 VI; same site, Tsagan-Gol river near somon, along bank, 3 VII 1945—Yun.; same site, Ikhe-Bogdo mountain range, plateau-like mountain crest, alpine carpet, 29 VII 1948—Grub.).

IB. Kashgar: ***Nor.*** (Bai district, No. 8237, 7 XI 1958—Lee and Chu); ***West.*** (Kingtau mountain range, 3–4 km from Koshkulak settlement, upper part of forest belt, 10 VI 1959—Yun.); ***East.*** (Luitsichen in Khomote, in Bagrashkul' lake region, 3100 m, No. 7185, 10 VIII 1958—Lee and Chu).

IIA. Junggar. ***Altay Region*** (Qinhe [Chingil'] river, 2630 m, No. 1054, 4 VIII; in Koktogoi region, No. 1942, 17 VIII 1956—Ching; in Timulbakhana region, on south. slope, 2450 m, Nos. 10700 and 10708, 18 and 19 VII 1959—Lee et al.); ***Tarb.*** (Saur mountain range, south. slope in Khobuk river valley, Karagaitu river gorge, right bank Bain-Tsagan creek, alpine belt, kobresia meadow, 23 VI; same site, subalpine meadow, 23 VI; same site, right-bank Bain-Gol creek, subalpine meadow, 23 VI 1957—Yun.; nor. of Chuguchak [Dachen], 2250 m, No. 1569, 13 VIII; Chuguchak, 1770 m, No. 1702, 14 VIII 1957—Kuan); ***Jung. Alt.*** (Kasan pass, 10 VIII 1878—A. Reg.; Toli district, No. 1197, 16 VIII and No. 2109, 26 VIII 1957—Kuan; 15 km nor.-west of Ven'tsyuan', 29 VIII 1957—Yun.); ***Tien Shan*** (several).

IIB. Tibet: ***Weitzan*** (on Dyaochu river, 11 VII 1884—Przew.).

IIIC. Pamir (Billuli river near its inflow into Chumbus river, on clayey-sandy-rocky soil 11 VI 1909—Divn.; Kokat pass at Yazag and Balung river sources, 3200–4000 m, 16 VI; Pil'nen gorge, 4500–5000 m, 1 VII; moraine water divide between Atrakyr and Tyuzutek rivers, moss tundra, 4500–5000 m, 20 VII 1942—Serp.).

General Distribution: Jung.-Tarb., Nor. and Cent. Tien Shan, East. Pam.; Near East (Afghanistan), Mid. Asia (West. Tien Shan, Pamiroalay), West. Sib. (Altay), East. Sib. (south.), North Mong. (Fore Hubs., Hent., Hang.), China (Altay, South-west.—nor. Sichuan), Himalayas (west., Kashmir).

Note. This Central Asian plant is very close to the Caucasian *C. tristis* M.B. and, in all probability, is a subspecies of the latter. *C. stenocarpa* exhibits much variability in number of staminate spikelets (1 to 3), arrangement of flowers in spikelets (plants with androgynous spikelets are found rather often), extent of pubescence of utricles (from rather pubescent in the upper part to altogether glabrous), width and consistency of leaves (3–6 mm broad), ratio of leaf to stem length (in alpine plants, leaves are half or more shorter than stem while in forest plants, the two are nearly equal) and in the ratio of utricle to glume length.

Section Capillares (Aschers et Graebn.) Rouy

68. **C. ledebouriana** C. A. Mey. ex Trev. in Bull. Soc. natur. Moscou, 36, 1 (1863) 540; V. Krecz. in Fl. SSSR, 3 (1935) 431; Grubov, Consp. fl. MNR (1955) 85; Fl. Kazakhst. 2 (1958) 79; Egorova in Nov. sist. vyssh. rast. (1964) 43. —*C. capillaris* var. *ledebouriana* (C. A. Mey. ex Trev.) Kük. in Pflanzenreich [IV, 20], 38 (1909) 591; Krylov, Fl. Zap. Sib. 3 (1929) 511. —*C. ruthenica* auct. non V. Krecz.: Grubov, l.c. 87, p.p., quoad pl. chobd.

Described from Altay. Type in Leningrad.

In wet meadows, steppe slopes, less often in forest; in alpine belt and upper forest belt.

IA. Mongolia: ***Khobd.*** (Katu river, on wet meadow soil, 16 VI; Bairimen-daban peak, 20 VI 1879—Pot.; Tolbo-Kungei Alatau mountain range, kobresia meadow, in high-altitude zone, 3200 m, 5 VIII; same site, alpine steppe, 5 VIII 1945—Yun.); ***Mong. Alt.*** (south. slope of Mongol. Altay, 25–30 km south of Tamchi-daba pass, midcourse of Bidzhi-Gol river, birch grove near spring source, 10 VII 1947—Yun.).

General distribution: Arct., Europe (Polar region and Nor. Urals), West. Sib. (Altay), East. Sib., Far East, North Mong. (Fore Hubs., Hent., Hang.).

69. **C. karoi** (Freyn) Freyn in Österr. Bot. Z. 46 (1896) 132; V. Krecz. in Fl. SSSR, 3 (1935) 430; Kitag. Lin. Fl. Mansh. (1939) 103; Fl. Kirgiz. 2 (1950) 300; Grubov, Consp. fl. MNR (1955) 85; Fl. Kazakhst. 2 (1958) 79; Fl. Tadzh. 2 (1963) 142; Egorova in Nov. sist. vyssh. rast. (1964) 40. —C.

capillaris ssp. *karoi* Freyn in Österr. Bot. Z. 40 (1890) 303. —*C. chamissonis* Meinsh. in Acta Horti Petrop. 18 (1901) 361; V. Krecz., l.c. 340. —*C. capillaris* f. *major* auct. non Drej.: Kük. in Pflanzenreich [IV, 20], 38 (1909) 590, p.p.; Krylov, Fl. Zap. Sib. 3 (1929) 511. —*C. capillaris* auct. non L.: Grubov, l.c. 82, p.p. —Ic.: Fl. Tadzh. 2, Plate XXXIII, figs. 5–8.

Described from East. Siberia. Type in Brno.

In wet and marshy meadows, in short-grass meadows near brooks, along borders of irrigation ditches, in scrub, in lower and middle hill zones (1800–2550 m above sea level).

IA. Mongolia: ***Khobd.*** (descent from Kharkhirinsk massif into Achit-Nur lake basin, short-grass meadow irrigated by brooks, 8 IX 1931—Bar.); ***Cis-Hing.*** (near Yaksha railway station, dry meadow, 11 VI 1902—Litv.); ***Cent. Khalkha, East. Mong., Depr. Lakes, Gobi-Alt.*** (Baga-Bogdo, in gravelly places, stream flats, 1800 m, No. 238, 1925—Chaney; Dzun-Saikhan hills, along banks of brooks, at top of Yalo creek, 23 VIII 1931—Ik.-Gal.).

IB. Kashgar: ***Nor.*** (Chzhaosu district in Aksu region, on slope, No. 3498, 14 VIII 1957—Kuan).

IIA. Junggar: ***Jung. Alt.*** (Upper Borotala, 2550 m, VIII 1878 A. Reg.); ***Tien Shan*** (midcourse of Taldy river, 2100 m, 26 V; Naryn-Gol river near its inflow into Tsagan-usu, 1800–2400 m, 10 VI; Mengute, 2700 m, 2 VIII 1879—A. Reg.).

IIIA. Qinghai: ***Amdo*** (Mudzhik river, along borders of irrigation ditches, 16 VI 1880—Przew.).

General distribution: Jung.-Tarb., Nor. and Cent. Tien Shan. East. Pam.; Mid Asia (West. Tien Shan, Pamiroalay), West. Sib. (south.), East Sib., Far East (Sea of Okhotsk coast), North Mong. (Fore Hubs., Hent., Hang., Mong.-Daur.), China (North), Himalayas (west., Kashmir), North Amer.

70. **C. sedakowii** C. A. Mey. ex Meinsh. in Acta Horti Petrop. 18 (1901) 360; Kük. in Pflanzenreich [IV, 20], 38 (1909) 588; Krylov, Fl. Zap. Sib. 3 (1929) 510; V. Krecz. in Fl. SSSR, 3 (1935) 427; Kitag. Lin. Fl. Mansh. (1939) 111; Grubov, Consp. fl. MNR (1955) 88. —*C. chaffanjonii* Camus in Lecomte, Not. Syst. Paris, 2, 7 (1912) 205; Danguy in Bull. Mus. nat. hist. natur. 20 (1914) 144; Grubov, Consp. fl. MNR (1955) 82. —**Ic.:** Russk. bot. zh. 3–6 (1911), fig. 129.

Described from East. Siberia (Transbaikal). Type in Leningrad.

IA. Mongolia: ***East. Mong.*** ("Vallée de la Kéroulen, steppe, alt. 900 m, 3 VI 1896, Chaff."—Camus, l.c.; Danguy, l.c.).

General distribution: West. Sib. (south., very rare), East. Sib., Far East. North Mong. (Fore Hubs., Hent., Mong.-Daur), China (Dunbei), Korea (nor.).

71. **C. selengensis** Ivanova in Bot. mater. Gerb. Bot. inst. AN SSSR, 7 (1937) 276; Egorova in Nov. sist. vyssh. rast. (1964) 42. —*C. capillaris* auct. non L.: Grubov, Consp. fl. MNR (1955) 82, p.p.

Described from North Mongolia. Type in Leningrad.

In wet meadows and in scrub.

IA. Mongolia: ***East. Mong.*** (near Kharkhonte railway station, in wet scrub, 8 and 21 VI 1902—Litw.).

General distribution: East. Sib. (south. parts of Buryatian Autonomous Soviet Socialistic Republic and Chita district), North Mong. (Hang., Mong.-Daur.), China (North).

Section Flavae (Carey) Christ

72. **C. aspratilis** V. Krecz. in Fl. SSSR, 3 (1935) 618 and 398; Grubov, Consp. fl. MNR (1955) 82; Egorova in Bot. mat. Gerb. Bot. inst. AN SSSR, 19 (1959) 80. —*C. diluta* auct. non M.B.: Kük. in Pflanzenreich [IV, 20], 38 (1909) 659, p.p., quoad pl. sibir.; Krylov, Fl. Zap. Sib. 3 (1929) 515, p. min. p. —**Ic.:** Fl. SSSR, 3, Plate XXI, fig. 7.

Described from Kazakhstan. Type in Leningrad.

In wet and marshy sites.

IA. Mongolia: ***Depr. Lakes*** (Ulangom, on marsh, 2 VII 1879—Pot.).

General distribution: Europe (south. half of Europ. USSR), West. Sib. (excluding Altay), East. Sib. (south.), North Mong. (Mong.-Daur.).

Note. The cited specimen of *C. aspratilis* differs from the type in weakly rugose beak of utricles without or with only stray spinules.

73. **C. diluta** M.B. Fl. taur.-cauc. 2 (1808) 388 and 3 (1819) 614; Kük. in Pflanzenreich [IV, 20], 38 (1909) 659, p.p.; Krylov, Fl. Zap. Sib. 3 (1929) 515, p.p.; Pampanini, Fl. Carac. (1930) 85; V. Krecz. in Fl. SSSR, 3 (1935) 394; Fl. Kirgiz. 2 (1950) 295; Fl. Kazakhst. 2 (1958) 74; Fl. Tadzh. 2 (1963) 130. —*C. karelinii* Meinsh. in Acta Horti. Petrop. 18 (1901) 380; V. Krecz. l.c. 395; Fl. Kazakhst. 2 (1958) 74; Egorova in Bot. mat. Gerb. Bot. inst. AN SSSR, 19 (1959) 81. —*C. chorgosica* Meinsh. l.c. 381. —*C. binervis* Meinsh. l.c. 383, p.p. —*C. diluta* var. *chorgosica* (Meinsh.) Kük. l.c. 660. —C. dibita var. karelinii (Meinsh.) Kük. l.c. 660. —**Ic.:** Fl. SSSR, 3, Plate XXI, fig. 8.

Described from the Caucasus. Type in Leningrad.

In wet rather saline meadows, along borders of irrigation ditches, near mouths of springs; from foothills to middle hill belt.

IA. Mongolia: ***Depr. Lakes*** ("Ulangom, wet rather saline meadows"—Egorova, l.c.).

IB. Kashgar: ***Nor.*** (Uchturfan, Yaman-su, along irrigation ditch, 9 VI 1908—Divn.); ***West.*** (valley of Gez-Darya river 100 km from Kashgar on road to Tashkurgan, near spring, 2420 m, 15 VI 1959—Yun.).

IIA. Junggar: ***Tien Shan*** (nor. of Kul'dzha, 3 V 1877; Cent. Khorgos, 900–1500 m, 15 VI 1878; Piluchi river gorge, 900–1500 m, 24 IV 1879—A. Reg.); ***Jung. Gobi*** (Sygashu, 600 m, 4 V 1879—A. Reg.; Usu [Shikho] district, San'tszyaochzhuan village, on wet site, 25 VI 1957—Kuan); ***Zaisan*** (Kaba river near Kaba village, in tugai, 16 VI 1914—Schischk.); ***Dzhark.*** (Kul'dzha, 5 V; Khoyur-Sumun south of Kul'dzha, 540 m, 26 V 1877—A. Reg.).

IIIC. Pamir. (Bulunkul', 2800 m, 15 VI 1959—Lee et al.).

General distribution: Aral-Casp., Balkh. Region, Jung.-Tarb., Nor. and Cent. Tien Shan; Europe (Europ. USSR, predominantly in south. half), Asia Minor, Near East, Caucasus, Mid. Asia, West. Sib. (south., including Altay where it is very rare), East. Sib. (south., except Sayan range), Himalayas (Kashmir).

Note. *C. karelinii* Meinsh. described from Balkhash Region (Lepsa river) is identical to *C. diluta*. The distinctive features of the latter species given by V. Kreczetowicz in "Flora of the USSR" apply in equal measure to both species. Meinshausen distinguished the species described by him from *C. distans* L. and *C. extensa* Good.

74. **C. philocrena** V. Krecz. in Fl. SSSR, 3 (1935) 618, 393; Fl. Kirgiz, 2 (1950) 294; Fl. Kazakhst. 2 (1958) 74; Fl. Tadzh. 2 (1963) 129. —*C. oederi*

auct. non Retz.: Kük. in Pflanzenreich [IV, 20], 38 (1909) 673, quoad pl. turkest. et songor. **Ic.:** Fl. Tadzh. 2, Plate XXX, figs. 1–3.

Described from Mid. Asia (Pamiroalay). Type in Leningrad.

In wet sites; in lower hill belt.

IIA. Junggar: ***Tien Shan*** (Dzhagastai, 600–900 m, 6 VIII 1877—A. Reg.; Nilki, VI 1879—A. Reg.; Tekes river, near river, in wet site, 10 VI 1893—Rob.).

General distribution: Jung.-Tarb., North and Cent. Tien Shan; Afghanistan, Mid. Asia (West. Tien Shan, Pamiroalay), China (North-west.—Gansu).

Note. C. *philocrena* is exceptionally close to the European species C. *oederi* Retz.

Section Secalinae O. Lang

75. **C. eremopyroides** V. Krecz. in Fl. SSSR, 3 (1935) 617, 384; Kitag. Lin. Fl. Mansh. (1939) 101; Grubov, Consp. fl. MNR (1955) 84. —*C. secalina* auct. non Wahl.: Kük. in Pflanzenreich [IV, 20], 38 (1909) 681, p.p., quoad pl. sibir. orient.

Described from East. Sib. Type in Leningrad.

In wet sandy locations.

IA. Mongolia: ***East. Mong.*** (Kulun-Buirnorskaya plain, Borol'd-zhitu, 26 VI 1899—Pot. and Sold.; near Manchuria railway station in wet location, No. 671, 11 VI; Khuna province, Sinbaerkhuyuchi district, on bank of Dalai lake, Nos. 975 and 989, 29 VI 1951, Li, S.H. et al.).

General distribution: East. Sib. (south. excluding Sayan range, very rare), China (Dunbei).

Note. This species is very close to C. *secalina* Wild. ex Wahl. growing in the eastern part of West. Europe, in the southern half of Europ. USSR, in the Caucasus, Kazakhstan, in the southern part of West. Sib. and in Minusinsk region. Almost all of the characteristics with which V. Kreczetowicz distinguishes his C. *eremopyroides* from C. *secalina* (short peduncle of staminate spikelets, different shape and smaller size of utricles, different ratio of length of utricles and glumes) were not confirmed when studying the herbarium specimens of C. *eremopyroides*.

The only difference between the species is that the utricles usually have (1) 2 or 3 veins on the back in C. *secalina* which, in the case of C. *eremopyroides*, are absent or single. This difference, too, is not however, absolute, since plants of C. *secalina* are encountered with utricles having veins on the back. The independent status of species C. *eremopyroides* calls for confirmation. It is possible that when a large number of herbarium specimens of C. *eremopyroides* become available (at present, not more than 15 are known), no differences may be discerned between these species with reference to venation.

Section Physocarpae (Drej.) Kük.

76. **C. dichroa** Freyn in Österr. Bot. Z. 40 (1890) 304; V. Krecz. in Fl. SSSR, 3 (1935) 447, p.p., excl. pl. Asiae centr.; Grubov, Consp. fl. MNR (1955) 83; Fl. Kazakhst. 2 (1958) 80.—*C. vesicaria* var. *grahamii* (Boott) Kük. in Pflanzenreich [IV, 20], 38 (1909) 727, p.p., quoad pl. sibir. —*C. kryloviana* Schischk. et Serg. in Sist. zam. Gerb. Tomsk. univ., 6 (1928) 1; Krylov in Fl. Zap. Sib. 3 (1929) 526. —*C. vesicaria* var. *alpigena* auct. non Fries; Kük. l.c. 727, quoad pl. Asiae. —**Ic.:** Fl. SSSR, 3 Plate XXIII, fig. 10.

Described from East. Sib. (Irkutsk district). Type in Vienna (?).

In marshy meadows; from alpine belt to upper forest limit.

IA. Mongolia: ***Khobd., Mong. Alt.*** (Sar-byuira, on bank of Dain-Gol lake, 25 IX 1876—Pot.; Bulgan somon, upper Indertiin-Gol, marshy meadow in high-altitude belt, 24 VII 1947—Yun.); ***Depr. lakes*** (Ulangom, in marsh, 3 VII 1879—Pot.; nor. bank of Ubsa lake, marshy forest, 3 VII 1892—Kryl.; 40 km south of Ulyasutai on road to Tsagan-Ol, Shurugin-Gol area, bottomland deciduous forest and marshy meadow, 17 VIII 1947—Letov).

General distribution: Jung.-Tarb.; West. Sib. (Altay), East. Sib. (south.), North Mong. (Fore Hubs., Hent., Hang.).

77. **C. pamirensis** Clarke in Kew Bull. Add. ser. 8 (1908) 87; idem, Trav. Soc. natur. St.-Pétersb. 35, 3 (1906) 191, nom. nud.; V. Krecz. in Fl. SSSR, 3 (1935) 448; Fl. Kirgiz. 2 (1950) 300; Fl. Tadzh. 2 (1963) 139; Ikonnikov, Opred. rast. Pamira (Key to Plants of Pamir) (1963) 83. —*C. vesicaria* var. *pamirica* O. Fedtsch. in Acta Horti Petrop. 21, 3 (1903) 432. —*C. pamirica* (O. Fedtsch.) V. Fedtsch. in Acta Horti. Petrop. 28, 1 (1908) 70, 123. —*C. obscuriceps* Kük. var. *pamirica* (O. Fedtsch.) Kük. in Pflanzenreich [IV, 20] 38 (1909) 724. —*C. ampullacea* auct. non Good.: Duthie in Alcock, Rep. Pamir Commiss. (1898) 27.

Described from Pamir (Rang-Kul' lake). Type in Leningrad.

On banks of rivers and lakes, on sides of canals, marshes and marshy meadows; in high-altitude belt.

IIA. Junggar: ***Altay Region*** (Qinhe, No. 0987, 4 VIII; Qinhe, Chzhunkhaitsza, in water, Nos. 1374 and 1414, 6 VIII 1956—Ching); ***Tien Shan*** (Mengute, 2700 m, 2 VIII; Kunges river, 2700 m, 31 VIII 1879—A. Reg.; B. Yuldus, 2 VIII 1893—Rob.; high-altitude basin of B. Yuldus, 30–35 km south-west of Bain-Bulak settlement, nor. fringe of basin floor, marshy meadow, encircles canal to depth up to 1 m, forms thickets, 10 VIII; same site, marshy meadow, 10 VIII 1958—Yun., B. Yuldus, near water, 2460 m, No. 6474, 10 VIII 1958—Lee and Chu).

IIIC. Pamir ("In marshy ground along the Aksu river, 3900–4200 m, Nos 17769 and 17770, Alcock"—Duthie, l.c.).

General distribution: North and Centr. Tien Shan, East. Pam.; Near East, Mid. Asia (West. Tien Shan, Pamiroalay), Himalayas (west., Kashmir).

Note. The Central Asian *C. pamirensis* is very close to the Mongolian-East Sib. *C. dichroa* Freyn; it differs from it in longer and narrower glumes, generally equal utricles (not shorter), broader leaves and very large size of the plant as a whole. However, these differences are far from absolute, due to which the species affinity of the specimens is often established quite tentatively. Within the distribution range of *C. pamirensis* specimens that are indistinguishable from *C. dichroa* are found. Even specimens collected from the same geographic location (Tien Shan—Yunatov) reveal two forms, one the typical *C. pamirensis* and another that can be regarded wholly as *C. dichroa.*

In the distribution range of *C. dichroa,* plants largely similar to *C. pamirensis* are found. Thus, an analysis of the material leads to the conclusion that *C. pamirensis* and *C. dichroa* represent in principle the subspecies of the same species, viz. *C. dichroa* s.l.

78. **C. pseudocyperus** L. Sp. pl. (1753) 978; Kük. in Pflanzenreich [IV, 20], 38 (1909) 695, p.p.; Krylov, Fl. Zap. Sib. 3 (1929) 519; V. Krecz. in Fl. SSSR, 3 (1935) 460; Fl. Kirgiz. 2 (1950) 301; Fl. Kazakhst. 2 (1958) 82. —**Ic.:** Fl. SSSR, 3, Plate XXIV, fig. 8.

Described from Sweden. Type in London (Linn.).

In grassy-sedge marshes and in water.

IIA. Junggar: *Jung. Gobi* (south.: 2 km nor. of Kuitun, No. 402, 6 VII 1957—Kuan; nor.-east of St. Kuitun settlement on Shikho-Manas highway, sasa zone, marshy meadow fed by spring, in water, 30 VI; same site, 3–4 km nor. of St. Kuitun settlement, sasa zone, grassy-sedge bog, 6 VII 1957—Yun.).

General distribution: Aral-Casp., Balkh. Region; Europe, Caucasus, Mid. Asia (West. Tien Shan), West. Sib. (excluding Altay), East. Sib. (south., excluding Sayan range), Himalayas (Kashmir—after Kükenthal, l.c.), Japan (Hokkaido island), North Amer., Afr. (north).

79. **C. rhynchophysa** C. A. Mey. in Suppl. Index Semin. Hort. Bot. Petrop. 9 (Jan.–Feb. 1844) 9; Danguy in Bull. Mus. nat. hist. natur. 20 (1914) 145; V. Krecz. in Fl. SSSR, 3 (1935) 440; Kitag. Lin. Fl. Mansh. (1939) 109; Grubov, Consp. fl. MNR (1955) 87. —*C. laevirostris* (Blytt ex Fries) Fries in Bot. Notis. 1–2 (Mar. 1844) 24; Kük. in Pflanzenreich [IV, 20], 38 (1909) 724; Krylov, Fl. Zap. Sib. 3 (1929) 523. —**Ic.:** Fl. SSSR, 3 Plate XXIII, fig. 4.

Described from specimens grown from seeds collected in Dauria. Type in Leningrad.

Along grassy-sedge marshes, marshy banks of rivers and brooks, marshy meadows.

IA. Mongolia: *Khobd.* (Grubov, l.c.), *Cis-Hing.* (left bank of Khalkhin-Gol stream in Numurgin-Gol upper region, herb meadow, 6–7 VIII 1949—Yun.; Arshan town, in wet meadow, No. 310, 13 VI 1950—Chang; "Kinghans, No. 1393, 27 VI 1896, Chaff."—Danguy, l.c.); *East. Mong., Depr. Lakes* (Ulangom, on banks of brook, on sandy soil, 26 VI; same site, in marsh, 2 VII; same site, near brook, 6 IX 1879—Pot.; near Ulangom, river bog, 1 IX 1930—Bar.).

IIA. Junggar: *Altay Region* (Qinhe, Chzhunkhaitsza, No. 1392, 6 VIII 1956—Ching).

General distribution: Arct., Europe, West. Sib., East. Sib. (excl. Sayan range), Far East, North Mong. (Hent., Hang., Mong.-Daur.), China (Dunbei), Korea (nor.), Japan, North Amer. (Canada).

80. **C. rostrata** Stokes in Wither. Arrang. Brit. Pl. ed. 2, 2 (1787) 1059; Kük. in Pflanzenreich [IV, 20], 38 (1909) 720, p.p.; Krylov, Fl. Zap. Sib. 3 (1929) 521. —*C. inflata* auct. non Huds.: V. Krecz. in Fl. SSSR, 3 (1935) 442; Grubov, Consp. fl. MNR (1955) 85; Fl. Kazakhst. 2 (1958) 80. —**Ic.:** Fl. SSSR, 3, Plate XXIII, fig. 3 (as *C. inflata*).

Described from England. Type not preserved.

In marshy meadows and in water.

IA. Mongolia. *Cent. Khalkha, East. Mong., Depr. Lakes* (Kharkhiry river stream 3–4 km south of Ulangom, marshy sedge meadows, 28 VII 1945—Yun.).

IIA. Junggar: *Altay Region* (Qinhe, Chzhunkhaitsza, in water, Nos. 1391 and 1412, 6 VIII 1956—Ching).

General distribution: Aral-Casp., Balkh. Region; Arct., Europe, Caucasus, West. Sib., East. Sib., Far East. Nor. Mong. (Fore Hubs., Hang.), China (Dunbei—B. Hinggan range), Nor. Amer.

81. **C. vesicata** Meinsh. in Acta Hort. Petrop. 18, 3 (1901) 367; V. Krecz. in Fl. SSSR, 3 (1935) 446; Kitag. Lin. Fl. Mansh. (1939) 113; Grubov, Consp. fl. MNR (1955) 89. —*C. vesicaria* var. *tenuistachya* Kük. in Bot. Centralbl. 77 (1899) 58; idem, Pflanzenreich [IV, 20], 38 (1909) 726. —*C. vesicaria* f. *tenuistachya* (Kük.) T. Koyama in J. Fac. Sci. Univ. Tokyo, sect. 3, 8 (1962)

242. —*C. vesicaria* auct. non L.: Forbes and Hemsley, Index Fl. Sin. 3 (1905) 317. —**Ic.:** Fl. SSSR, 3, Plate XXIII, fig. 2.

Described from the Far East (Amur river). Type in Leningrad.

In wet meadows, along river banks.

IA. Mongolia: ***Cis-Hing.*** (left bank of Khalkhin-Gol stream in upper Numurgin-Gol region, herb meadow, 6–7 VIII 1949—Yun.; Yaksha railway station, hummocky marsh, No. 2225, 1954—Wang (Khalkha-Gol somon, Khamar-Dabana region, floodplain of river, 19 VI 1954—Dashnyam). ***Khobd., Cent. Khalkha.***

IIA. Junggar: ***Jung.-Gobi.***

General distribution: Arct. (Asiat.), East. Sib., Far East, North Mong. (all districts), China (Dunbei), Korea (nor.), Japan.

Section Carex

82. **C. drymophila** Turcz. ex Steud. Syn. Cyper. (1855) 238; Kük. in Pflanzenreich [IV, 20] 38 (1909) 755, p.p.; Danguy in Bull. Mus. nat. hist. natur. 20 (1914) 144; V. Krecz. in Fl. SSSR, 3 (1935) 456; Kitag. Lin. Fl. Mansh. (1939) 100. —**Ic.:** Fl. SSSR, 3, Plate XXIV, fig. 4.

Described from East. Siberia (Transbaikal). Type in Leningrad.

In banks of water reservoirs.

IA. Mongolia: ***Cis-Hing.*** ("Kinghans, vallée du Djatan-Gol, No. 1387, 28 VI 1896, Chaff." Danguy, l.c.).

General distribution: East. Sib., Far East, China (Dunbei, Nor.) Korea.

83. **C. orthostachys** C. A. Mey. in Ledeb. Fl. Alt. 4 (1833) 231; Franch. Pl. David. 1 (1884) 321; V. Krecz. in Fl. SSSR, 3 (1935) 452; Kitag. Lin. Fl. Mansh. (1939) 107; Grubov, Consp. fl. MNR (1955) 86; Fl. Kazakhst. 2 (1958) 82. —*C. aristata* var. *orthostachys* (C. A. Mey.) Clarke in Forbes and Hemsley, Index Fl. Sin. 3 (1905) 272, pp. —C. aristata ssp. orthostachys (C.A. Mey.) Kük. in Pflanzenreich [IV, 20], 38 (1909) 753; Krylov, Fl. Zap. Sib. 3 (1929) 534. —*C. raddei* auct. non Kük.: Egorova in Bot. mat. Gerb. Bot. inst. AN SSSR, 19 (1959) 81. —**Ic.:** Fl. SSSR, 3, Plate XXIV, fig. 2.

Described from Kazakhstan (Zaisan lake). Type in Leningrad.

In wet and marshy meadows, along banks of rivers.

IA. Mongolia: ***Cis-Hing.*** (Yaksha railway station, in a meadow on roadside, No. 2268, 1954—Wang); ***Mong. Alt., Cent. Khalkha, East. Mong.*** (near Kharkhonte railway station, basin between sand-hills, 7 VI 1902; same site, wet meadow, 15 VI 1903—Litw.; Khalkha-Gol somon, Khamar-Daban region, carex meadow in floodplain of river, 19 VIII 1954; Bayan-Ula somon, valley bottom, carex marshy meadow, 31 VII 1956—Dashnyam); ***Val. Lakes.***

General distribution: Aral-Casp., Balkh. Region; Europe (south. Urals), West. Sib. (south.), East. Sib. (south.), Far East (south.), North Mong. (Fore Hubs., Hent., Hang., Mong.-Daur.), China (Dunbei, North).

84. **C. raddei** Kük. in Bot. Centralbl. 77 (1899) 97; V. Krecz. in Fl. SSSR, 3 (1935) 457; Kitag. Lin. Fl. Mansh. (1939) 108. —*C. aristata* ssp. raddei (Kük.) Kük. in Pflanzenreich [IV, 20], 38 (1909) 755. —**Ic.:** Fl. SSSR, 3, Plate XXIV, fig. 3.

Described from the Far East (Zeya mouth). Type (?). Isotype in Leningrad.

IA. Mongolia: ***Cis-Hing.*** (near Yaksha railway station, in thick grass, 11 VI 1902—Litw.).
General distribution: Far East (south.), China (Dunbei), Korea (nor.).

Section Paludosae (Fries) Christ

85. **C. acutiformis** Ehrh. Beitr. Naturk. Wissensch. 4 (1789) 43; Kük. in Pflanzenreich [IV, 20], 38 (1909) 733; Krylov, Fl. Zap. Sib. 3 (1929) 527; V. Krecz. in Fl. SSSR, 3 (1935) 400; Fl. Kirgiz. 2 (1950) 295; Fl. Kazakhst. 2 (1958) 75; Fl. Tadzh. 2 (1963) 132. —*C. paludosa* Good. in Trans. Linn. Soc. London, 2 (1794) 202. **Ic.:** Fl. Tadzh. 2, Plate XXXI, figs. 1–3.

Described from West. Europe, Type in Leningrad.

IIA. Junggar: ***Tien Shan*** (Mengute, 2700 m, 2 VIII 1879—A. Reg.; in Shikhetsza region, No. 1169, 2 VII 1957—Kuan).

General distribution: Aral-Casp., Balkh. Region, Cent. Tien Shan; Europe, Asian Minor, Near East (Iraq, Syria), Caucasus, Mid. Asia (except Turkmenia), West. Sib. (south., excluding Altay), East. Sib. (in Minusinsk and Achinsk regions), ? Himalayas (east., Kashmir), Afr. (Algeria).

86. **C. gotoi** Ohwi in Mem. Coll. Sci. Kyoto Univ. ser. B, 5 (1930) 248; Kitag. Lin. Fl. Mansh. (1939) 102. —*C. sukaczevii* V. Krecz. in Fl. Zabaik. 2 (1931) 136; idem, Fl. SSSR, 3 (1935) 415. —*C. haematostachys* auct. non Lévl. et Vaniot: V. Krecz. l.c. 1935, 415. —*C. heterostachya* auct. non Bunge: Kük. in Pflanzenreich [IV, 20], 38 (1909) 741, p.p. —*C. nutans* auct. non Host: Danguy in Bull. Mus. nat hist. natur. 20 (1914) 144. —*C. songorica* auct. non Kar. et Kir.: Egorova in Bot. mater. Gerb. Bot. inst. AN SSSR, 19 (1959) 81, p.p., quoad pl. chingan.

Described from Korean peninsula. Type in Kyoto.

IA. Mongolia: ***Cis-Hing.*** ("Vallée du Djatangol, steppes, alt. 800 m, 27 VI 1986, Chaff."—Danguy, l.c.; near Yaksha railway station, steppe, 11 VI 1902—Litw.; same site, dry slope, No. 2186, 1954—Wang; 5 km west of Toge-Gol river, grass meadow, 7 VIII 1949—Yun.; Khalkha-Gol somon, Dege and Numurgina interfluve, grass meadow, 28 VIII 1963—Dashnyam).

General distribution: East. Sib. (south., excluding Sayan range), Far East (south.), North Mong. (Mong.-Daur)., China (Dunbei), Korea, Japan.

***C. heterostachya** Bunge in Mèm. Ac. Sc. St.-Pètersb. Sav. Étr. 2 (1832) 142; Forbes and Hemsley, Index Fl. Sin. 3 (1905) 289; Kük. in Pflanzenreich [IV, 20], 38 (1909) 741, p.p.; Kitag. Lin. Fl. Mansh. (1939) 102; Chen and Chou, Rast. pokrov r. Sulekhe (Vegetational Cover of Sulekhe River) (1957) 81; Koyama in J. Fac. Sc. Univ. Tokyo, sect, 3, 8, 4 (1962) 247. —*C. nutans* auct. non Host.: Franch. Pl. David. 1 (1884) 321; ?Forbes and Hemsley, Index Fl. Sin. 3 (1905) 301.

Described from North China. Type in Paris.

IA. Mongolia: ***Khesi*** ("Sulekhe river"—Chen and Chou, l.c.).
General distribution: China (Dunbei, North and North-west.), Korea.

87. **C. melanostachya** M.B. ex Willd. Sp. pl. 4 (1805) 299; V. Krecz. in Fl. SSSR, 3 (1935) 412; Fl. Kirgiz. 2 (1950) 296; Fl. Kazakhst. 2 (1958) 76; Fl. Tadzh. 2 (1963) 136. —*C. nutans* Host, Gram. Austr. 1 (1801) 61, non J.F. Gmel. (1791); ?Forbes and Hemsley, Index Fl. Sin. 3 (1905) 301; Kük. in Pflanzenreich [IV, 20], 38 (1909) 740, p.p.; Krylov, Fl. Zap. Sib. 3 (1929) 530. —*C. pumila* var. *nutans* (Host) T. Koyama in J. Fac. Sc. Univ. Tokyo, sect. 3, 8, 4 (1962) 248. —**Ic.:** Fl. SSSR, 3 (1935), Plate XXII, fig. 6.

Described from the Caucasus. Type in Berlin.

IIA. Junggar: ***Dzhark.*** (Chimpanzi village, 8 V; Piluchi, 11 V; Khoyur-Sumun, 24 V 1877—A. Reg.).

General distribution: Aral-Casp., Balkh. Region, Jung.-Tarb., Nor. and Cent. Tien Shan; Europe (Cent. Europe and south. half of Europ. USSR), Mediterr., Balk.-Asia Minor, Near East (Iran, ?Afghanistan), Caucasus, Mid. Asia, West. Sib. (south., excluding Altay), China (North and North-west.).

Note. References to the occurrence of *C. melanostachya* (= *C. nutans* Host) in North. and Nor.-west. China (Forbes and Hemsley, Kükenthal, l.c.) in all probability pertain to *C. heterosachya* Bunge. The two species are externally similar.

88. **C. riparia** Curt. Fl. Lond. (1777–1798) Tab. 60; Kük. in Pflanzenreich [IV, 20], 38 (1909) 735, p.p.; Krylov, Fl. Zap. Sib. 3 (1929) 529; V. Krecz. in Fl. SSSR, 3 (1935) 409; Fl. Kirgiz. 2 (1950) 295; Fl. Kazakhst. 2 (1958) 75; Fl. Tadzh. 2 (1963) 134. —**Ic.:** Curt. l.c. Tab. 281; Fl. SSSR, 3, Plate XXII, fig. 1.

Described from England. Type in London (K).

Along river banks.

IIA. Junggar: ***Tien Shan*** (Dzhagastai, 1500–1800 m, 8 VIII 1877—A. Reg.).

General distribution: Aral-Casp., Balkh. Region, Nor. and Cent. Tien Shan; Europe, Asia Minor, Near East, Caucasus, Mid. Asia (West. Tien Shan and ?Pamiroalay), West. Sib. (south., excluding Altay), East. Sib. Minusinsk region), Afr. (north).

89. **C. rugulosa** Kük. in Bull. Hérb. Boiss., 2 sér. 4 (1904) 58; V. Krecz. in Fl. SSSR, 3 (1935) 410; Kitag. Lin, Fl. Mansh. (1939) 110. —*C. riparia* var. *rugulosa* Kük. in Pflanzenreich [IV, 20], 38 (1909) 736. —*C. smirnovii* V. Krecz. l.c. 619, 411.

Described from Japan. Type in Leningrad.

Along river banks.

IA. Mongolia: ***Khesi*** (Gaotai, in Kheikho river valley, 20 VI; Khechen village, 27 VI 1886—Pot.).

General distribution: West. Sib. (excluding Altay, very rare), East. Sib. (south., excluding Sayan range), Far East (south.), Japan.

Note. *C. smirnovii* V. Krecz., described from East. Siberia, was later identified by the author as *C. rugulosa* Kük., as can be seen from his later identification of herbarium specimens of this species. *C. rugulosa* is very close to *C. riparia* Curt. The difference between the species is that the glumes in *C. riparia* are longer than the utricles and have a very long awn, while in *C. rugulosa,* they are shorter than the utricles and with/without very short awn and acute. Moreover, the leaves of *C. riparia* are usually broader than those of *C. rugulosa.* These species do not differ in utricles.

90. **C. songorica** Kar. et Kir. in Bull. Soc. natur. Mosc. 15 (1842) 525; V. Krecz. in Fl. SSSR, 3 (1935) 414; Fl. Kirgiz. 2 (1950) 299; Grubov, Consp. fl. MNR (1955) 88; Fl. Kazakhst. 2 (1958) 76; Fl. Tadzh. 2 (1963) 137; Ikonnikov, Opred. rast. Pamira (Key to Plants of Pamir) (1963) 83. —*C. heterostachya* auct. non Bunge; Kük. in Pflanzenreich [IV, 20], 38 (1909) 741, p.p.; Krylov, Fl. Zap. Sib. 3 (1929) 531. —**Ic.:** Fl. SSSR, Plate XXII, fig. 10.

Described from Kazakhstan (Lepsa river). Type in Leningrad.

In wet and marshy, often rather saline and solonchak meadows fed by spring, along banks of rivers and lakes, less often in larch forests; predominantly from submontane plains to lower and middle mountain belts, very rarely ascending even into high altitudes (up to 2700 m above sea level in Tien Shan).

IA. Mongolia: ***Depr. Lakes*** (Ulangom, in marsh, 2 VII; same site, on clayey dry soil, 3 VII 1879—Pot.; valley of Sagli river, near spring mouths, 21 IX 1931—Bar.).

IIA. Junggar: ***Altay Region*** (Qinhe [Chingil'], ravine floor, No. 113, 8 VII; same site, along hill slope, No. 846, 2 VIII; between Ertai and Qinhe, in gorge, No. 737, 29 VII 1956—Ching); ***Tarb.*** (Tumandy river, 7 VIII 1876—Pot.; valley of Khobuk river, larch forest, 20 VII 1914—Sap.); ***Jung. Alt.*** (Borotaly river upper course, 1800 m, VIII 1878—A. Reg.; Toli region, in gorge, No. 1141, 7 VIII; same site, near water, No. 2755, 8 VIII 1957—Kuan); ***Tien Shan, Jung. Gobi*** (south.: Sygashu, 4 VIII 1879—A. Reg.; (Bodonchi river, stream border, 19 VII 1947—Yun.) Usu district [Shikho], San'tszyaochzhuan village, in wet site, No. 1050, 25 VI; in Kuitun region, on meadow, No. 271, 29 VI; 3 km nor. of Kuitun, No. 363, 6 VII 1957—Kuan; 2–4 km nor.-east of St. Kuitun settlement, on Shikho-Manas highway, marshy meadows, 30 VI and 6 VII 1957—Yun.; east.: Uienchi somon, Borotsonchzhi area, solonchak carex-herb meadows, 13 IX 1948—Grub.).

General distribution: Aral-Casp., Balkh. Region, Jung-Tarb., Nor. and Cent. Tien Shan, East. Pam.; Near East Caucasus (South. Transcaucasus), Mid Asia, West. Sib. (Altay), East. Sib. (upper Yenisey basin), North Mong. (Mong.-Daur.).

Family **ARACEAE** Juss.

1. **Acorus** L.

Sp. pl. (1753) 924; idem, gen. pl., ed. 5 (1754) 151.

1. **A. calamus** L. Sp. pl. (1735) 324; Forbes and Hemsley, Index Fl. Sin. 3 (1905) 187; Krylov, Fl. Zap. Sib. 3 (1929) 538; Kuzen. in Fl. SSSR, 3 (1935) 479; Grubov, Consp. fl. MNR (1955) 89; Fl. Kazakhst. 2 (1958) 83.

Described from Europe. Type in London (Linn.).

On river banks, near water, in marshes.

IA. Mongolia: ***Cis-Hung.*** (near Yaksha railway station, meadow marsh, 19 VIII 1902—Litw.); ***East. Mong.*** (Durbetsi-beise railway station, 13 VII 1909—Rudnev; Khailar town, Nunlin'tun' village, near water, 10 VI 1951—Wang).

General distribution: Jung.-Tarb.; Europe, Asia Minor, ?Near East, Caucasus, West. Sib. (south., excluding Altay), East. Sib., Far East (south.), North Mong. (Mong.-Daur.), China (Dunbei, North, North-west., Cent., East.), Himalayas (west., east.), ?Korea, Japan, India, North Amer.

Family LEMNACEAE Gray

1. Lemna L.

Sp. pl. (1753) 970; idem, gen. pl., ed. 5 (1754) 417.

1. Stems oblong or lanceolate, slender, semitransparent, altogether flat .... 2. **L. trisulca** L.
+ Stems orbicular or broadly elliptical, or ovate, thickened, spongy, not transparent, slightly convex above and flat below .... 1. **L. minor** L.

1. **L. minor** L. Sp. pl. (1753) 970; Forbes and Hemsley, Index Fl. Sin. 3 (1905) 188; Krylov, Fl. Zap. Sib. 3 (1929) 540; Kuzen. in Fl. SSSR, 3 (1935) 493; Persson in Bot. Notis. (1938) 277; Kitag. Lin. Fl. Mansh. (1939) 125; Fl. Kirgiz. 3 (1951) 11; Grubov, Consp. Fl. MNR (1955) 89; Fl. Kazakhst. 2 (1958) 87; Fl. Tadzh. 2 (1963) 165. —**Ic.:** Fl. Tadzh. 2, Plate XXXVIII, figs. 2 and 3.

Described from Europe. Type in London (Linn.).

In standing and slow-moving water: small pools in meadows, rather saline lowlands, grass plots, marshes, irrigation ditches, rice fields.

IA. Mongolia: ***Cis-Hing.*** (near Yaksha railway station, meadow lake, 19 VIII 1902—Litw.); ***Cent. Khalkha*** (Borokhchin-Gol brook, 3 IX 1926—Prokhanov; south. border of Tsagan-Nur lake lowland on Ulan Bator-Tsetserleg road, small fresh water lakelets in sands, 25 VI 1948—Grub.), ***East. Mong., Val. Lakes,*** (Orok Nor, surface of lagoon with algae, at 1135 m, No. 319, 1925—Chaney); ***Depr. Lakes*** ("Khorgon-Shibir"—Grubov, l.c.); ***Gobi-Alt.*** (Bain-Tukhum area, meadow solonchaks, in water, 4 VIII 1931—Ik.-Gal.; same site, 20–25 km west of Bain-Dalai somon, rather saline lowland, VII-VIII 1933—Simukova); ***Ordos*** (along Ulan-Morin river, in standing pool, 23 VIII; along Baga-Gol river, marshy banks, 13 IX 1884—Pot.).

IB. Kashgar: ***West.*** (near Yangishar, in marsh, 25 VII 1929—Pop.; "Kashgar, in rice fields, about 1330 m, 8 V 1934"—Persson, l.c.); ***East.*** (Khami, in standing water near irrigation ditch, 4 VI 1877—Pot.).

IIA. Junggar: ***Jung. Alt.*** (Ven'tsyuan', Nos. 4578 and 4591, 23 and 24 VIII 1957—Kuan); ***Jung. Gobi*** (nor.: Urungu river, in stagnant pool, 21 VIII 1876—Pot.).

General distribution: Aral-Casp., Balkh. Region; Arct. (Europ., extremely rare), Europe, Asia Minor, Near East, Caucasus, Mid. Asia (West. Pamiroalay, rare, West. Tien Shan), West. Sib. (excluding Altay), East. Sib. (excluding Sayan range), Far East, North Mong. (Hang., Mong.-Daur.), China (Dunbei, North, South.-west.), Korea, Japan, Indo-Malay., North Amer., Afr. (North), Austral. Almost cosmopolitan.

2. **L. trisulca** L. Sp. pl. (1753) 970; Krylov, Fl. Zap. Sib. 3 (1929) 540; Kuzen. in Fl. SSSR, 3 (1935) 493; Kitag. Lin. Fl. Mansh. (1939) 125; Fl. Kirgiz. 3 (1951) 11; Grubov, Consp. fl. MNR (1955). 89; Fl. Kazakhst. 2 (1958) 86; Fl. Tadzh. 2 (1963) 155. —**Ic.:** Fl. Tadzh. 2, Plate XXXVIII, fig. 1.

Described from Europe. Type in London (Linn.).

In stagnant water.

IA. Mongolia. ***East. Mong., Val. Lakes, Ordos*** (along Naryn-Gol river, in stagnant water, 11 and 12 IX 1884—Pot.).

IIA. Junggar: ***Tien Shan*** (Ulumbai oasis 20–25 km south of Urumchi, in water, together with green mosses, 2 VI 1957—Yun.).

General distribution: Balkh. Region, Nor. Tien Shan; Arct. (rare), Europe, Asia Minor, Near East, Caucasus, Mid. Asia (West. Pamiroalay—Gissar), West. Sib., East. Sib., Far East, North Mong. (Fore Hubs., Hent., Hang., Mong.-Daur), China (Dunbei), Korea, Japan, North Amer., Afr., Austral.

Family **JUNCACEAE** Juss.

1. Leaf sheaths closed, with fascicle of hairs in aperture; leaf margin generally with rather long hairs. Fruit—unilocular capsule with three seeds...2. **Luzula** DC.
+ Leaf sheaths open; leaves glabrous. Fruit—trilocular many-seeded capsule ....1. **Juncus** L.

1. **Juncus** L.

Sp. pl. (1753) 325; idem, gen. pl., ed. 5 (1754) 152.

1. Plants annual...2.
+ Plants perennial ....4.
2. Perianth segments 3–4 mm long, largely equal. Capsule spherical ....3. **J. sphaerocarpus** Nees.
+ Perianth segments 6–7.5 mm long, outer ones much longer than inner...3.
3. Inflorescence diffuse, divergent; flowers usually single or in twos ....1. **J. bufonius** L.
+ Inflorescence compressed, with contorted branches; flowers in twos and threes....................2. **J. nastanthus** V. Krecz. et Gontsch.
4. Stem bases surrounded by scaly sheaths; leaf blades absent or only most upper sheath with very short (not more than 2 cm long) blade. Inflorescence apparently lateral since bract leaf present at its base appears as a direct extension of stem ....5.
+ Stems foliate, sheaths with leaf blades. Inflorescence not lateral..... 7.
5. Inflorescence with not more than 10 flowers. Stems slender, about 1 mm in diameter..23. **J. filiformis** L.
+ Inflorescence many-flowered. Stems 2–4 mm in diameter .... 6.
6. Stems strongly grooved, easily flattened, inside with transverse septums of spongy tissue. Capsule acuminate at top..................... ....22. **J. brachytepalus** V. Krecz et Gontsch.
+ Stems very weakly and finely grooved, not flattened, inside wholly filled with spongy tissue, without septums. Capsule flat at top ....***J. decipiens** (Buchenau) Nakai, p. 113.

7 (4). Flowers aggregated in one head, surrounded by general involucre of bracts and appearing like one flower. Anthers as long as filaments or shorter; stamens often exserted from perianth. Seed with long white tail-like appendages at both ends...............................8.
+ Inflorescence composite, sometimes capitate, but then anthers 3–5 times longer than barely developed filaments; stamens shorter than perianth or as long, not exserted; seed without tail-like appendages..14.
8. All leaves radical ....9.

+ Apart from radical, one cauline leaf present, separated from radical leaf by aphyllous stem part ....11.

9. The lowest bract leaf-like, up to 3.5 cm long. Perianth segments white ....5. **J. leucomelas** Royle.

+ All bracts scale-like, boat-shaped. Perianth segments from whitish to purplish brown ....10.

10. Anthers about 0.7 mm long; stamens shorter than perianth segments. Bracts appressed to inflorescence ....12. **J. triglumis** L.

+ Anthers 1.8–2.5 mm long; stamens longer than perianth, exserted from it. Bracts horizontally deflexed ......10. **J. thomsonii** Buchenau.

11 (8). Perianth segments purplish brown .....8. **J. przewalskii** Buchenau.

+ Perianth segments white or yellowish-white ....12.

12. Perianth segments 4–5 mm long, equal to stamens or slightly longer. Inflorescence with 1 or 2, rarely 3 flowers........................ ....7. **J. potaninii** Buchenau.

+ Perianth segments 6–7.5 mm long shorter than stamens (stamens prominent above flowers). Inflorescence many-flowered .... 13.

13. Leaves, especially radical, appear as though articulate owing to prominent transverse septums ....4. **J. allioides** Franch.

+ Leaves without prominent transverse septums................................ ....6. **J. membranaceus** Royle.

14(7) Leaves flatly-cylindrical, inside with transverse septums of spongy tissue, extremely prominent in dry state (leaf as though articulate owing) ....15.

+ Leaves not cylindrical, without, septums flat or grooved-convoluted ....19.

15. Perianth segments wholly dark brown, without light-coloured veins and white membranous margin, finely acuminate. Capsule abruptly contracted into long beak ....20. **J. atratus** Krock.

+ Perianth segments not wholly coloured; midrib light-coloured or with white membranous margin or greenish; if brown all over, not acuminate but capsule with short beak ....16.

16. Perianth segments brown or purplish-brown. Capsule brown or chestnut-brown, elliptical or oblong-elliptical, abruptly contracted into short beak ....17.

+ Perianth segments pale green or pale brownish. Capsule greenish, trigonous-conical, gradually narrowed into beak ....*J. **leschenaultii** J. Gay, p. 112.

17. Perianth segments (2.5) 3–3.5 mm long, outer ones with white membranous margin, acute, but without spinule. Capsule oblong-elliptical, 3.5–4 mm long....................................19. **J. articulatus** L.

+ Perianth segments 2–2.5 mm long, outer ones usually without white membranous margin, obtuse, with short spinule. Capsule elliptical, 2.5–3 mm long..18.

18. Perianth segments equal. Anthers nearly as long as filaments or slightly shorter ....18. **J. alpinus** Vill.
+ Perianth segments equal or inner slightly longer than outer ones. Anthers 1/2 to 2/3 as long as filaments.. ....21. **J. turczaninowii** (Buchenau) V. Krecz.

19 (14) Stems and bract leaves not more than 1 mm in diameter. Capsule 2–5 mm long. Seeds nut-like, without white tail-like appendages. Flowers generally sessile and single on branches, rarely aggregated in heads ....20.
+ Stems thickened, 2–3 mm in diameter; lower bract leaf 3–5 mm broad at base. Capsule 7.5–10 mm long. Seeds with long tail-like appendages. Flowers aggregated in heads, forming umbellate inflorescence ....24.

20. Inflorescence corymbose-paniculate, lax. Capsule 2.5–3.2 mm long ....21.
+ Inflorescence aggregated, compact, with 1 to 5 heads. Capsule 3.5–5 mm long...23.

21. Flowers single at ends of branches, rarely in twos ....22.
+ Flowers in fascicles of 2 or 3 at ends of branches............................ ....17. **J. soranthus** Schrenk.

22. Capsule subspherical, orbicular on top, usually double the size of perianth segments. Bracts white-membranous. Anthers almost as long as filaments...13. **J. compressus** Jacq.
+ Capsule obovate or ovate, equal to or slightly longer than perianth segments. Bracts rusty-brown, sometimes pale but not white. Anthers 3–4 times longer than filaments................14. **J. gerardii** Lois.

23. Capsule considerably longer than perianth segments....................... ....15. **J. heptapotamicus** V. Krecz. et Gontsch.
+ Capsule equal to or slightly shorter than perianth pinnae................. ....16. **J. salsuginosus** Turcz.

24 (19) Perianth segments 8–9 mm long, almost as long as capsule. Capsule abruptly contracted into beak.............9. **J. sphacelatus** Decne.
+ Perianth segments about 5 mm long, almost half shorter than capsule. Capsule gradually narrowed into beak...........11. **J. tibeticus** Egor.

Subgenus **Tenageia** (Dum.) Kuntze

1. **J. bufonius** L. Sp. pl. (1753) 328; Franch. Pl. David. 1 (1884) 311; Forbes and Hemsley, Index Fl. Sin. 3 (1905) 162; Buchenau in Pflanzenreich [IV, 36], 25 (1906) 105; Krylov, Fl. Zap. Sib. 3 (1929) 561; Pampanini, Fl. Carac. (1930) 86; V. Krecz. and Gontsch. in Fl. SSSR, 3 (1935) 517; Persson in Bot. Notis. (1938) 277; Kitag. Lin. Fl. Mansh. (1939) 127; Walker in Contribs. U.S. Nat. Herb. 28 (1941) 600; Grubov, Consp. fl. MNR (1955) 90; Fl.

Tadzh. 2 (1963) 163. —? *J. erythropodus* V. Krecz. in Byull. Sredneaz. gos. univ., 21 (1935) 176; Fl. Kirgiz. 3 (1951) 13; Fl. Kazakhst. 2 (1958) 90.

Described from Europe. Type in London (Linn.).

Along wet and highly saturated sites: silty and sandy banks of rivers and pebble beds, near springs and brooks, in shallow water, in saline meadows and solonchaks, along irrigation ditches; from foothills to high altitudes.

IA. Mongolia: ***Mong. Alt., Cis-Hing.*** (Yaksha railway station, in a meadow, No. 2839, 1954—Wang); ***Cent. Khalkha*** (Ubur-Dzhargalante river, near Dol'che-Gechen monastery, 14 VIII 1925—Krasch. and Zam.; ear Ikhe-tukhum-Nor lake (46°5′ N lat., 104–105° E long.); Ata-Bulak sources, VI 1926—Zam.; Tsetsen-Khanovskii khoshun, valley of Kerulen river, 6 VIII 1927—Zam.; 5–7 km east of Choiren-ul, saline meadow, 23 VIII 1940—Yun.); ***East. Mong., Depr. Lakes*** (near Khirgiz-Nur lake, in wet sandy sites, 7 VIII 1879—Pot.; Shargin-Gobi, bank of Shargin-Gol river, solonchaks, 8 IX 1930—Pob.), ***Val. Lakes, Gobi-Alt.*** (Naryn-Bulak sources, 16 VIII 1886—Pot.; Bain-Tukhum area, solonchaks, 4 VIII 1931—Ik.-Gal.; same site, rather saline lowland, VII–VIII 1933—Sim.); ***East. Gobi*** (Khushu-khid monastery on Ongin-Gole river Floodplain) 22 X 1947—Grub. and Kal.); ***West. Gobi*** (south. foot of Tsagan-Bogdo mountain range, rather saline meadow, 1 VIII; Bain-Gobi somon, Tsagan-Burgasun area, solonchaks and rather saline meadows, 8 VIII 1943—Yun.); ***Alash. Gobi*** (Edzin-Gol river, Dzhargalante area, 17 VI 1909—Czet.; valley of Edzin-Gol river near upper Ontsin-Gol, in a water-filled pit, 13 VII 1926—Glag.; 40 km nor. of Yunchan town, meadow in valley, 1 VII 1958—Petr.); ***Ordos*** (Ulan-Morin, 23 VIII; Dzhasygen area, 30 VIII 1884—Pot.; 25 km south-east of Otok town, rather saline meadow near Khaolaitunao lake, 1 VIII; 20 km west of Dzhasak town, meadow, 16 VIII 1957—Petr.); ***Khesi*** (Yanchi village, 29, VI 1886—Pot.; "Liu Fu Jai, near streams, on frequently flooded, sandy and gravelly soil and on steppes, 1923, Ching"—Walker, l.c.).

IB. Kashgar: ***Nor.*** (Yarkand-Darya, along river streams 900 m, 22 VI 1889—Rob.; Uchturfan, Yaman-su, along sasa, 9 VI 1908—Divn.; Pichan district, Lyamusin, near water, No. 6659, 13 VI 1958—Lee and Chu); ***West.*** ("Bostan-terek, about 2400 m, 6 VIII 1934"—Persson, l.c.); ***South.*** (Niya oasis, 1260 m, 3 VI 1885—Przew.); ***East.*** (Khami, 14 VI 1879—Przew.).

IIA. Junggar: ***Altay Region*** (Qinhe [Chingil'], 1700 m, No. 0881, 2 VIII; Fuyun'-Ukagou, No. 1699, 11 VIII 1956—Ching); ***Jung. Alt.*** (Toli-Myaergou, No. 2421, 4 VIII 1957—Kuan); ***Tien Shan*** (Piluchi, 20 VI 1877—A. Reg.; Algoi, 1800–2400 m, 12 IX 1879—A. Reg.); ***Jung. Gobi*** (south.: Savan district, Katszyvan, No. 1315, 9 VII; Gunlyu-Shakhe [Shikho], No. 3653, 18 VIII 1957—Kuan; east.: Baitak-Bogdo mountain range, Takhiltu-ula, in water, 17 IX 1948—Grub.); ***Zaisan*** (right bank of Ch. Irtysh below Burchum river, 15 VI 1914—Schischk.).

IIIA. Qinghai: ***Nanshan*** (in wooded central part of mountain range falling south of Tetung river, 2550 m, 4 VIII 1880—Przew.; South-Kukunorsk mountain range, Usubin-Gol, 3150 m, 16 VIII 1901—Lad.).

IIIB. Tibet: ***Weitzan*** (nor. slope of Burkhan-Budda mountain range, Khatu gorge, 3150 m, 21 VII 1901—Lad.

General distribution: Aral-Casp., Balkh. Region, Jung.-Tarb., Nor. and Cent. Tien Shan; Arct. (stray finds), Europe, Asia Minor, Near East, Caucasus, West. Sib., East. Sib., Far East, North Mong., China, Himalayas, Korea, Japan, ?Indo-Malay., North Amer., Afr. (nor.); introduced into South Amer. and Austral.

Note. Variable plant providing grounds for taxonomists to establish many taxa from and within it. In particular, Middle and Central Asian representatives of *J. bufonius* L. were described by V. Kreczetowicz (l.c.) as *J. erythropodus* V. Krecz. on the basis of their inner perianth segments being equal to fruit length, but not exceeding it and, moreover, perianth slightly shorter than in *J. bufonius* s. str. In Central Asian plants of the type of *J. bufonius* studied by us, the length ratio of fruit and inner perianth segments varies: the latter are either as long as the fruit or longer. A very similar variation is also noticed even among European material of *J. bufonius*. Thus, the independent status of plants placed under *J. erythropodus* is, in our opinion, highly dubious.

Another species closely related to *J. bufonius*, viz. *J. turkestanicus* V. Krecz. et Gontsch., has been cited for Kul'dzha by Kreczetowicz and Gontscharov in "Flora of the USSR". The Herbarium of the Botanical Institute, Academy of Sciences, USSR (Leningrad), however, does not contain speciemens of species from this region.

2. **J. nastanthus** V. Krecz. et Gontsch. in Fl. SSSR, 3 (1935) 517; Fl. Kirgiz. 3 (1951) 13; Grubov, Consp. fl. MNR (1955) 91; Fl. Kazakhst. 2 (1958) 92; Fl. Tadzh. 2 (1963) 166. —**Ic.**: Fl. SSSR, 3, Plate XXX, fig. 18.

Described from Europe (Leningrad district). Type in Leningrad.

IA. Mongolia: ***Mong. Alt.*** (nor.-west.—Grubov, l.c.); ***Depr. Lakes.***

IIA. Junggar: ***Tien Shan*** (Dzhagastai, 600–900 m, 6 VIII 1877—A. Reg.).

General distribution: Aral-Casp., Balkh. Region, ?North Tien Shan, Europe, Caucasus, Mid. Asia (West. Tien Shan, West. Pamiroalay), West. Sib., East. Sib.

Note. This species representing in principle one of the variants of *J. bufonius*, is highly dubious.

3. **J. sphaerocarpus** Nees in Flora, 1 (1818) 521; Buchenau in Pflanzenreich [IV, 36], 25 (1906) 108; Krylov, Fl. Zap. Sib. 3 (1929) 562; V. Krecz. and Gontsch. in Fl. SSSR, 3 (1935) 516; Fl. Kazakhst. 2 (1958) 90; Fl. Tadzh. 2 (1963) 162. —**Ic.**: Fl. SSSR, 3, Plate XXX, fig. 17.

Described from Europe. Type in London (K).

IIA. Junggar: ***Jung. Gobi*** (nor.: along Ch. Irtysh river, 26 VIII 1876—Pot.).

General distribution: Aral-Casp. (nor.), Balkh. Region; Europe, Asia Minor, Caucasus, Mid. Asia (West. Pamiroalay and Turkmenia—Ashkhabad region), West. Sib. (south., excluding Altay).

Subgenus **Stygiopsis** (Gand.) Kuntze

4. **J. allioides** Franch. in Nouv. Arch. Mus. hist. natur. 10 (1887) 99; Forbes and Hemsley, Index Fl. Sin. 3 (1905) 162; Buchenau in Pflanzenreich [IV, 36], 25 (1906) 229; Walker in Contribs. U.S. Nat. Herb. 28 (1941) 600. —*J. macranthus* Buchenau, Monogr. Juncac. (1890) 398. —**Ic.**: Pflanzenreich [IV, 36], 25, fig. 106.

Described from China (Dunbei). Type in Paris. Plate VIII, fig. 8.

Along wet and marshy localities; in alpine zone, less often in forest zone.

IIIA. Qinghai. ***Nanshan*** (on nor. slope of South Tetungsk mountain range, in alpine zone, among shrubs, 1 VIII; same site, in forest belt, 6 VIII 1880—Przew.; "T'ai Hua (Ping Fan Hsien), on a moist densely bushy mountain top and in a partially shaded swamp, forming dense stands, No. 507, 1923, Ching"—Walker, l.c.); ***Amdo*** (along Mudzhik river, near spring, among shrubs, 2700–2800 m, 16 VI 1880—Przew.).

General distribution: China (North-west., Cent., South-west.).

5. **J. leucomelas** Royle in Trans. Linn. Soc. London, 18, 3 (1840) 320; Buchenau in Pflanzenreich [IV, 36], 25 (1906) 225; Pampanini, Fl. Carac. (1930) 86.

Described from north-western Himalayas. Type in London (K). Plate VIII, fig. 5.

In high-altitude zone, along river banks.

IIIA. Qinghai: ***Nanshan*** (mountain range along Tetung river, in alpine belt, VI 1872—Przew.).

IIIB. Tibet: ***Chang Tang*** ("Lago Pancong, coll. Schlagintweit"—Pampanini, l.c.); ***Weitzan*** (Yangtze river basin, upper course of Ichu river, 3900 m, 29 VII 1901—Lad.).

General distribution: China (South-west.), Himalayas.

6. **J. membranaceus** Royle in Trans. Linn. Soc. London, 18, 3 (1840) 317; Deasy, In Tibet a. Chin. Turk. (1901) 404; Buchenau in Pflanzenreich [IV, 36], 25 (1906) 229; Pampanini, Fl. Carac. (1930) 87.

Described from Himalayas. Type in London (K).

IIIB. Tibet: ***Chang Tang*** ("Aksu, 4740 m, 1898"—Deasy, l.c.).

General distribution: Himalayas.

7. **J. potaninii** Buchenau, Monogr. Juncac. (1890) 394; Forbes and Hemsley, Index Fl. Sin. 3 (105) 165; Walker in Contribs. U.S. Nat. Herb. 28 (1941) 602. —*J. luzuliformis* var. *potaninii* (Buchenau) Buchenau in Pflanzenreich [IV, 36], 25 (1906) 228. —**Ic.:** Pflanzenreich [IV, 36], 25, fig. 105.

Described from China (Gansu province). Lectotype in Leningrad.

IIIA. Qinghai: ***Nanshan*** (in midportion of forest zone of South Tetungsk mountain range, 2500 m, 4 VIII 1880—Przew.; "La Chi'iung Kou, in tufts, on dénsely shaded, rocky cliffs by a stream, common 1923, Ching"—Walker, l.c.).

General distribution: China (Nor.-west. south-west).

8. **J. przewalskii** Buchenau, Monogr. Juncac. (1890) 401; idem, Pflanzenreich [IV, 36], 25 (1906) 231. —**Ic.:** Pflanzenreich [IV, 36], 25, fig. 108.

Described from China (Gansu). Type in Leningrad. Plate VIII, fig. 4.

In alpine belt.

IIIA. Qinghai: ***Nanshan*** (along Tetung river, in alpine belt, VII 1872—Przew., type!).

General distribution: China (South-west.).

9. **J. sphacelatus** Decne. in Jacquemont, Voy. l'Inde (1844) 172; Buchenau in Pflanzenreich [IV, 36, 25 (1906) 233, p.p., excel. pl. turkest.; Strachey, Catal. (1906) 192; ?Pampanini, Fl. Carac. (1930) 87. —**Ic.:** Pflanzenreich [IV, 36], 25, fig. 110.

Described from Himalayas. Type in ?Paris.

IIIB. Tibet: ***Weitzan*** (water divide between Yangtze and Mekong rivers, along Gochu brook, 3900 m, 23 VIII 1900; nor. slope of Burkhan-Buda mountain range, Khatu gorge, 3150 m, 17 VI 1901—Lad.).

General distribution: ?Near East (Afghanistan), China (South-west.), Himalayas.

10. **J. thomsonii** Buchenau in Bot. Zeit. 25 (1867) 148; idem, Pflanzenreich [IV, 36], 25 (1906) 224; Deasy, In Tibet a. Chin. Turk. (1901) 398; Hemsley, Fl. Tibet (1902) 200; Forbes and Hemsley, Index Fl. Sin. 3 (1905) 166; Strachey, Catal. (1906) 192; Ostenfeld in Hedin, S. Tibet (1922) 90; Rehder in J. Arn. Arb. 14 (1933) 5; V. Krecz. and Gontsch. in Fl. SSSR, 3 (1935) 523; Fl. Kirgiz. 3 (1951) 18; Fl. Tadzh. 2 (1963) 167; Ikonnikov, Opred.

rast. Pamira (Key to Plants of Pamir) (1963) 85. —*J. membranaceus* auct. non Royle: Hemsley in J. Linn. Soc. (London) Bot. 30 (1894) 119.

Described from nor.-west. Himalayas. Type? Plate VIII, fig. 6.

Along banks of rivers and brooks, near springs, in marshy and wet short-grass meadows; in alpine belt, 2700–5500 m above sea level.

IIIA. Qinghai: ***Nanshan*** (South Tetungsk mountain range, Myn'dan'sha river, 24 V 1890—Gr.-Grzh.; Sharagol'dzhin river, Buklu-Tologoi area, in wet meadow, 3000 m, 13 VI 1894—Rob.); ***Amdo*** (along Baga-Gorgi river, near springs, 2700 m, 25 V; along Mudzhik river, 2700–2850 m, 16 VI 1880—Przew.).

IIIB. Tibet: ***Chang Tang*** ("Camp 26, 34°38′ N lat., 82°10′ E long., 4830 m, 31 VII 1896"—Deasy, l.c.; "close to water, 4860 m, Thorold, 102"—Hemsley, l.c.; "S.W. Tibet, Tokchen, camp 211, 4654 m, 24 VII 1901, Hedin"—Ostendfeld, l.c.); ***Weitzan*** (left bank of Yangtze river, 3300 m, 24 VII 1884—Przew.; Mekong basin, 1900—Lad.; Amnen-kor mountain range, nor. slope, Kuku-Bulak spring, 3750 m, 5 VI 1900—Lad.; Alyk-Norin-Gol river valley, 3630 m, 7 VI 1901—Lad.); ***South.*** ("Eastern Tibet, Ba Valley, No. 14248, 1925–1926, Rock"—Rehder, l.c.).

IIIC. Pamir (Kenkol river, Tom-Kara area, along bank of brook, in marsh, 14 VI 1909—Divn.; near Takhtakorum pass, 4500–5500 m, 1 VIII 1942—Serp.; "Mustaghata, Kemper-kishlak, damp meadows beneath the glacier, about 4500 m, 29 VII 1894, Hedin"—Ostenfeld, l.c.).

General distribution: East. Pam.; Mid. Asia (Pamiroalay), China (Nor.-west., South-west.), Himalayas (west., Kashmir).

Note. Perianth segments vary in colour from purple to chestnut-brown to nearly white; often pale purple.

11. **J. tibeticus** Egor. sp. nov.—Planta perennis, 50–60 cm alt., rhizomate repenti, alte foliata. Folia inferiora subplana, superiora canaliculata. Inflorescentia ramosa, 3–10-capitata, capitulis 5-12-floris. Bractea infima frondosa inflorescentiam superans. Bracteae florum albineae vel pallide purpureo-fuscae vel pallide-roseae. Tepala aequalia ca. 5 mm lg., lanceolata acuta, pallide-castanea; stamina perigonia paulo breviore, antheris 0.7–1.2 mm lg. Capsula elongato-conica, convexo-trigona, ca. 7.5–8 mm lg., perigonium subduplo superans, in rostrum sensim angustata, seminibus 3.5 mm lg. appendiculis longis angustis caudiculatis instructis.

Typus: China, Nanschan, in regione sylvarum jugi a fl. Tetung S versus in latere boreali ad rivulum, 9 VII 1872, N.M. Przewalski. In Herb. Inst. Bot. Acad. Sci. URSS (Leningrad) conservatur.

Affinitas. A specie proxima *J. leucochlamis* Zing. ex V. Krecz. fructis elongato-conicis (nec cylindraceo-ellipticis) in rostrum sensim (nec abrupte) angustati differt.

Plate VIII, fig. 3.

In marshes and along banks of rivers and brooks; in forest and alpine zones.

IIIA. Qinghai: ***Nanshan*** (in forest belt of South Tetungsk mountain range, on nor. slope, along brook, 9 VII 1872—Przew., type!; mountain range north of Tetung river, in alpine zone, in marsh, 28 VIII 1872; South Tetungsk mountain range, 1 VIII; same site, forest zone, 6 VIII 1880—Przew.); ***Amdo*** (along Mudzhik river, 2700–2850 m, 16 VI 1880—Przew.).

General distribution: China (Nor.-west.).

Note. This species is well differentiated from the related *J. leucochlamis* Zing. ex V. Krecz.

distributed in the southern part of Eastern Siberia and in Northern Mongolia in oblong-conical (not cylindrical-elliptical) fruits, gradually narrowed (not abruptly) into beak without flexure.

12. **J. triglumis** L. Sp. pl. (1753) 328; Buchenau in Pflanzenreich [IV, 36], 25 (1906) 224; Krylov, Fl. Zap. Sib. 3 (1929) 576; V. Krecz. and Gontsch. in Fl. SSSR, 3 (1935) 522; Fl. Kirgiz. 3 (1951) 18; Grubov, Consp. fl. MNR (1955) 91; Fl. Kazakhst. 2 (1958) 93; Fl. Tadzh. 2 (1963) 166; Ikonnikov, Opred. rast. Pamira (Key to Plants of Pamir) (1963) 84. —*J. schischkinii* Kryl. et Sumn. in Sistem. zam. Gerb. Tomsk. univ., 7 (1928) 1; Krylov, Fl. Zap. Sib. 3 (1929) 579; V. Krecz. and Gontsch., l.c., 522; Kitag. Lin. Fl. Mansh. (1939) 128; Grubov, Consp. fl. MNR (1955) 91. —**Ic.:** Fl. Tadzh. 2, Plate XL, figs. 7 and 8.

Described from Europe (hills of Sweden). Type in London (Linn.). Plate VIII, fig. 7.

Along wet and marshy sites on banks of rivers and brooks, marshy meadows; predominantly in alpine zone, 2100–4500 m above sea level; less often, in upper forest boundary zone.

IA. Mongolia: ***Khobd.*** (Tszusylan, in forest, 13 VI; Kharkhira river below Tyurgun, on sandy soil, 21 VII 1879—Pot.); ***Mong. Alt.*** (Bulgan somon summer camp in upper Indertiin-Gol, marshy meadow in high-altitude zone, 24 VII 1947—Yun.); ***Depr. Lakes.***

IIA. Junggar: ***Jung. Alt.*** (between Sumbe and Kasan, 2100–2400 m, 22 VI 1878—A. Reg.); ***Tien Shan*** (Dzhagastai hill, 2400–2700 m, 20 VI 1878—A. Reg.; Dzhagastai-Gol, 2700 m, 5 IX; Aryslyn, 2400 and 2700–3000 m, 8 and 16 VII; Aryslyn-daban, 7 VIII; Mengute, 2700 m, 9 VII and 2 VIII 1879—A. Reg.; Passe zwischen Kinsu u. Kurdou, 3 VII 1907—Merzbacher; 7–8 km south of Dan, No. 522, 22 VII 1957—Kuan; left bank of Manas river, Danu-Gol river upper course, marshy sedge meadow, 22 VII 1957—Yun.; B. Yuldus river, in marsh, No. 6466, 10 VIII 1958—Lee and Chu); ***Jung. Gobi.***

IIIB. Tibet. ***Chang Tang*** (mountain range along Kyuk-Egil' river 3750–3900 m, 11 VII 1885—Przew.; left bank of Karakash river 15 km east of Kirgiz-dzhangil pass on Tibet road, river valley, rather saline meadow, 2 VI 1959—Yun.; Sinkiang-Tibet highway, 14–15 km east of Karakash pass, 4500 m, No. 00471, 2 VI 1959—Lee et al.).

General distribution: Jung.-Tarb., Nor. and Cent. Tien Shan, East. Pam; Arct., Europe (hills of Scandinavia and Central Europe, Nor. Urals), Caucasus, Mid. Asia (West. Tien Shan, Pamiroalay), West. Sib. (Altay), East. Sib. (Sayan range) North Mong., Himalayas.

Note. *J. schischkini* Kryl. et Sumn. described from Altay and treated here as a synonym of *J. triglumis* L. was differentiated from the latter in white perianth segments, of which the inner are shorter than the outer, and also the fruit surpasses the perianth only slightly. Our investigations showed that *J. triglumis* varies in colour of the perianth segments throughout the distribution range from purplish-brown to white with wide-ranging colour transitions. Further, it has not been possible to establish the relationship between colour of perianth and ratio between its length and that of fruit. Irrespective of the colour of the perianth, the fruit surpasses the perianth by almost twice or one-third or is almost equal to it. Similar colour variation of the perianth has also been observed among *J. thomsonil* Buchenau, which is closer to *J. triglumis*. Insofar as the length variation between outer and inner perianth segments is concerned, this is not a specific feature of *J. schischkinii*; it is characteristic of *J. triglumis* as well.

It is possible that reports on the occurrence of *J. triglumis* in the Himalayas pertain to *J. thomsonii*.

Subgenus **Pseudotenageia** V. Krecz. et Gontsch.

13. **J. compressus** Jacq. Enum. stirp. Vindob. (1762) 60, 235; Franch, Pl. David. 1 (1884) 311; Buchenau in Pflanzenreich [IV, 36], 25 (1906) 111; Krylov, Fl. Zap. Sib. 3 (1929) 565; V. Krecz. and Gontsch. in Fl. SSSR, 3 (1935) 527; Grubov, Consp. fl. MNR (1955) 90; Fl. Kazakhst. 2 (1958) 94; Fl. Tadzh. 2 (1963) 170. —*J. gracillimus* (Buchenau) V. Krecz. et Gontsch. in Flore SSSR, 3 (1935) 528; Kitag. Lin. Fl. Mansh. (1939) 127. **Ic.:** Fl. SSSR, 3, Plate XXVIII, fig. 1.

Described from Europe (Austria). Type ?

Along wet localities, marshy and floodplain meadows, water reservoirs, grassy marshes.

IA. Mongolia: ***Khobd.*** (left bank of Kharkhiry river, 9 VII 1879—Pot.); ***Mong. Alt.*** (Buluguna river valley, 18 IX 1930—Bar.); ***Cis-Hing.*** (Khalkhin-Gol river valley, 13 km south-east of Khamar-Daban,-willow thickets, 11 VIII 1949—Yun.; Arshan town near hot springs, 15 VI 1950—Chang; Yaksha railway station, No. 2818, 1954—Wang), ***Cent. Khalkha, East. Mong.*** (22 km east of Tumen-Delger somon, sedge-reed bog, 8 VIII 1956; Choibalsan somon, 6 km south of Enger-Shand, Kerulen river floodplain, in swampy sections, 18 VIII 1957—Dashnyam; Shilin-Khoto town, 1959—Ivan.); ***Depr. Lakes*** (around Ubsa lake, 3 VII 1892—Kryl.); ***Gobi-Alt.***

IIA. Junggar: ***Altay Region*** (west of Fuyun', No. 1894, 13 VIII Altay, No. 2731, 6 IX 1956—Ching); ***Tarb.*** (nor. of Dachen, No. 2942, 13 VIII 1957—Kuan); ***Jung. Gobi*** (nor.: Barbagai-Burchum, Nos. 2940 and 2998, 1 and 6 IX 1956—Ching; south.: near Takianzi, 300 m, 24 VIII 1878—A Reg.; in Chanzy region, No. 5027, 20 IX; Usu district, San'tszyao-chzhuan, No. 1049, 25 VI 1957—Kuan); ***Dzhark.*** (Kul'dzha, 1876—Golike; Khoyur-Sumun south of Kul'dzha, 27 V 1877—A. Reg.); ***Jung.-Gobi.***

General distribution: Jung.-Tarb.; Europe, Caucasus, Mid. Asia, West. Sib., East. Sib., Far East, North Mong. (Hang., Mong.-Daur.), China (Dunbei, North, Nor.-west.), ?Himalayas (Kashmir), Korea, Japan.

14. **J. gerardii** Lois. in J. Bot. 2 (1809) 284; Buchenau in Pflanzenreich [IV, 36], 25 (1906) 112, p.p.; Krylov, Fl. Zap. Sib. 3 (1929) 566; V. Krecz. and Gontsch. in Fl. SSSR, 3 (1935) 528; Persson in Bot. Notis. (1938) 277; Fl. Kirgiz. 3 (1951) 14; Grubov, Consp. fl. MNR (1955) 90; Fl. Kazakhst. 2 (1955) 94; Fl. Tadzh. 2 (1963) 171 —*J. atrofuscus* Rupr. Fl. samojed. (1845) 59; V. Krecz. and Gontsch. l.c. 529. —*J. gerardii* var. *acutiflorus* Buchenau, l.c. 113. —**Ic.:** Fl. SSSR, 3, Plate XXVIII, fig. 3.

Described from Europe (France). Type in Paris. Plate VIII, fig. 9.

In rather saline marshy meadows, solonchaks, grassy swamps, pebble beds and silty-sandy sites along river banks, near brooks; predominantly in submontane plain, in foothills and lower hill zone; rises in some regions up to 3100 m above sea level.

IA. Mongolia: ***Khobd.*** (Bekon-Beren river, Altyn-Khatysyn, in rather saline, sandy-silty sites, 18 VI; Burgasutai river in Uryuk-Nor lake basin, on pebble bed, 21 VI 1879—Pot.); ***Mong. Alt.*** (Tsagan-Derisu river valley, in marsh, 3 VII 1877—Pot.; Buyantu river valley, 28 VIII 1930—Bar.; Khasagtu-Khairkhan hills, Dundu-Tseren-Gol river, on pebble bed, 16 IX 1930—Pob.; Bodonchi river floodplain 2–3 km above Bodonchiin-khure, silted stream, 19 VII 1947—Yun.); ***East. Mong., Depr. Lakes*** (Ulangom, 2 VI 1879—Pot.; Shargin-Gobi, near Gol-ikhe, solonchaks, 4, IX 1930—Pob.); *Gobi-Alt.* (Dzhirgalante-Bulak spring, 23 VII 1926 and VIII. 1931—Ik.-Gal.); *West. Gobi* (Tsagan-Bogdo mountain range, Suchzhi-Bulak collective, saline meadow, 4 VIII; Tel' somon, Dzakhoi-Dzaram area south of Mong. Altay, rather saline basin, 18 VIII 1943—Yun.); *Alash. Gobi* (along Edzin river, Tsagan-beli, 16 VII 1886—Pot.); *Khesi* (Shakhe village, 4; between Gaotai and Fuiitin, 12 VI; near Goatai, in marsh, 20 VI 1886—Pot.).

IB. Kashgar: *Nor.* (Uchturfan, on sasa, 14 V 1908—Divn.; Pichan district, west of Khando lake, 460 m, near water, No. 6662, 13 VI 1958—Lee and Chu); *West.* (Artush-Khalatsi, No. 09780, 22 VI 1959—Lee et al.; "Jerzil, on fenny gravel soil, about 3100 m, 21 VII 1930"—Persson, l.c.; "Kashgar, 25 km south of the town in a swamp, about 1330 m, 12 V 1935"—Persson, l.c.); *East.* (Lyaodun' in Khami region, No. 452, 4 V 1957—Kuan).

IC. Qaidam: *Mount.* (Kurlyk-Nor and Toso-Nor lakes, 2580 m, 28 VI; Tuguryuk area near Barun khyrma, in saline marsh, 2580 m, No. 334, 1 VIII 1901—Lad.; in west up to Urtu-Bulak spring, in solonchaks, 2760 m, 10 IX 1884—Przew.).

IIA. Junggar: *Altay Region* (Qinhe, No. 0880, 2 VIII 1956—Ching; Koktogoi, near water, 950 m, 7 VI; same site, 900 m, 10 VI 1959—Lee et al.); *Tarb.* (Khobuk, No. 3410, 23 IX 1956—Ching); *Tien Shan* (nor. of Kul'dzha, 3 V; south. bank of Sairam lake, 1200 , 20 VII 1877—A. Reg.); *Jung. Gobi* (cent.: 83 km south of Ertai, on Altay-Guchen road, rather saline grass plot, 15 VII 1959—Yun.; west.: Darbuty river valley at its crossing on Karamai-Altay highway, saline reed thicket, 20 VI; 8–10 km south of Darbuty river, spring, 20 VI 1957—Yun.; south.: Savan district, Mogukhu water reservoir, No. 1565, 25 VI; in Shikhetsza region, No. 1173, 2 VI; Gunlyu-Shakhe [Shikho], No. 3654, 18 VII 1957—Kuan; east.: Uienchi somon, Borotsonchzihi area, saline meadow, 13 IX 1948—Grub.); *Dzhark.* (Kul'dzha, 5 V; along Ili river, 16 V; Sumun south of Kul'dzha, 29 V 1877—A. Reg.).

General distribution: Aral.-Casp., Balkh. Region, ?Jung.-Tarb.; Arct. (Europ.), Europe, Asia Minor, Near East, Caucasus, Mid. Asia (Turkmenia, West. Tien Shan and West. Pamiroalay), West. Sib. (south.), East. Sib., North Mong. (Hent., Hang., Mong.-Daur), ?China.

Note. Among plants described as *J. gerardii* var. *acutiflorus* Buchenau, l.c. the inner perianth segments are somewhat acute but not obtuse as in the type form. In the present case, however, acute tips of perianth segments represent the result of damage at the end of vegetation of their membranous broadly-orbicular margin. This variant has no taxonomic significance.

The species varies in colour intensity of the perianth throughout its distribution range. In our view, therefore, isolating *J. atrofuscus* Rupr. cannot be justified.

15. **J. heptapotamicus** V. Krecz. et Gontsch. in Fl. SSSR, 3 (1935) 628, 530; Fl. Kirgiz. 2 (1951) 17; Fl. Kazakhst. (1958) 95; Fl. Tadzh. 2 (1963) 172. —**Ic.:** Fl. SSSR, 3, Plate XXVIII, fig. 6.

Described from East. Kazakhstan. Type in Leningrad.

Near springs and brooks in swamps and marshy meadows; middle mountain zone.

IB. Kashgar: *Nor.* (Muzart river valley in upper course, Sazlik area, meadow, 9 IX 1958—Yun.); *West.* (Gez-Darya river valley, 100 km south of Kashgar on road to Tashkurgan, near spring, 15 VI 1959—Yun.).

IIA. Junggar: *Tien Shan* (Toguztorau near Kul'dzha, 5 V 1877—A. Reg.).

IIIC. Pamir (Chumbus river, Kosh-Terek area, along bank of brook, 8 V; Yangisarskii district, Egiz-yar village, along bank of brook, 8 V; Charlysh river, marsh on bank of brook, 21 VI 1909—Divn.).

General distribution: Balkh. Region, Jung.-Tarb., North and Centr. Tien Shan; Mid. Asia (Pamiroalay—Turkestan mountain range).

16. **J. salsuginosus** Turcz. in Bull. Soc. natur. Moscou, 28, 1 (1855) 304; V. Krecz. and Gontsch. in Fl. SSSR, 3 (1935) 530; Grubov, Consp. fl. MNR

(1955) 91. —*J. gerardii* var. *salsuginosus* (Turcz.) Buchenau in Pflanzenreich [IV, 36], 25 (1906) 113. —**Ic.:** Fl. SSSR, 3, Plate XXVIII, fig. 5.

Described from East. Siberia (Transbaikal). Type in ?Kharkov. Plate VIII, fig. 10.

In rather saline marshy meadows, solonchaks, in marshes, near springs.

IA. Mongolia: ***Khobd.*** (Altyn-Khatysyn, Bekon-Beren river, 18 VI 1879—Pot.); ***Mong. Alt.*** (2–3 km south-east of Yusun-Bulak, rather saline lowand, 13 VII 1947—Tuvanzhab; east. bank of Tonkhil'-Nur lake, meadow solonchaks, 16 VII; Tsitsiriin-Gol river upper course, creek floor, 24 VII 1947—Yun.); ***Cent. Khalkha*** (Ubur-Dzhargalante river, near Dol'che-Gegen monastery, meadow, 14 and 29 VIII; Ubur-Dzhargalante river mouth, swampy meadows, 10 IX; water divide between Ara-Dzhargalante and Ubur-Dzhargalante rivers, in marsh, 15 IX; Ubur-Dzhargalante river upper course, meadow, 25 IX 1925, Krasch. and Zam.; around Ikhe-Tukhum-Nur lake, trough at Ubur-bulgain-ama, 24 VII 1926—Zam.; Ulan-Bator—Tsetserleg road, sands around southern border of Tsagan-Nur lake, rather saline grass plot, 25 VI 1948—Yun.); ***Depr. Lakes*** (in meadows along Shuryk river, tributary of Dzabkhyn river, 23 VII 1877—Pot.; Baga-Nor lake, 30 VII; Gumburde area, 30 VII 1879—Pot.; nor. bank of Khirgiz-Nur lake, near irrigated plantations, 21 VIII 1944—Yun.); ***Val. Lakes, Gobi Alt.*** (Bain-Dalai somon, Bain-Tukhum area, VII–VIII 1943—Yun.; 1 km from Dalan-Dzadagad town, rather saline meadow, VII-VIII 1939—Surmazhab; 7 km south of Dzhinet somon, rather saline meadow, 30 VII 1941—Tsatsenkin; south. trail of Bain-Tsagen mountain range 10 km nor.-west of Bain-Bulak, 7 VII 1945—Yun.; south of Gurban-Saikhan mountain range, near Baishinte-Khure monastery, 29 VI 1946—Kondrat'ev; south. trail of Artsa-Bogdo range, Dzhirgalant-khuduk collective, 20 VII; Bain-Gobi somon, Tsagan-Gol river, near camp at somon, sedge meadow, 27 VII 1948—Grub.); ***East. Gobi*** (Grubov, l.c.); ***West. Gobi*** (south. foot of Tsagan-Bogdo mountain range, Tsagan-Bulak area, saline meadow near spring, 1 VIII 1943—Yun.).

General distribution: East. Sib. (Transbaikal), North Mong.

17. **J. soranthus** Schrenk in Bull. phys.-math. Ac. Sci. St.-Pétersb. 2 (1843) 194; V. Krecz. and Gontsch. in Fl. SSSR, 3 (1935) 531; Fl. Kazakhst. 2 (1958) 96. —**Ic.:** Fl. SSSR, 3, Plate XXVIII, fig. 8.

Described from Kazakhstan. Type in Leningrad.

In wet sandy sites, solonchak, near irrigation ditches, in scrub on banks.

IA. Mongolia: ***Mong.-Alt.*** (along Tsitsiriin-Gol river, on sandy wet sites, 10 VII 1877—Pot.).

IB. Kashgar: ***Nor.*** (1 km nor. of Pichan district, near water, No. 5430, 23 V 1958—Lee and Chu); ***East.*** (Turfan: Toksun district, 2 km east of Irakhu, meadow, No. 7271, 15 VI 1958—Lee and Chu).

IIA. Junggar: ***Tarb.*** (along Tumandy river, 8 VIII 1876—Pot.); ***Jung. Gobi*** (south.: Manas river basin, "30th Regiment" state farm, No. 1622, 9 VII 1957—Kuan; same site, 6–7 km nor. of "30th Regiment" state farm, near irrigation ditches, 9 VII 1957—Yun.; east.: on Barkul' lake, No. 4906, 25 IX 1957—Kuan); ***Zaisan*** (left bank of Ch. Irtysh river opposite Cherektas hill, solonchak, 11 VI; right bank of Ch. Irtysh river below Burchum river, willow groves, 15 VI 1914, Schischk.).

General distribution: Aral-Casp., Balkh. Region; Europ (south. Europ. USSR), Caucasus (Cis.-Caucasus), West. Sib. (south., excluding Altay), Mid. Asia (West. Tien Shan); *North Mong.* (Hang.).

Note. This species is very close to *J. gerardii* Jacq. Plants exhibiting the characteristics of *J. soranthus* should evidently be treated as a form of *J. gerardii.*

Subgenus **Ozophyllum** (Dum.) Kuntze

18. **J. alpinus** Vill. Hist. pl. Dauph. 2 (1787) 233; Buchenau in Pflanzenreich, [IV, 36], 25 (1906) 214; Krylov, Fl. Zap. Sib. 3 (1929) 575; V. Krecz. and Gontsch. in Fl. SSSR, 3 (1935) 537; Grubov, Consp. fl. MNR (1955) 90; Fl. Kazakhst. 2 (1958) 98. —**Ic.:** Fl. SSSR, 3, Plate XXX, fig. 4.

Described from France. Type in Paris.

In marshy meadows, in tugais.

IA. Mongolia: ***Khobd., Mong. Alt.*** Bulugun-Gol bend near Bulugun somon, bottomland poplar forest, 22 IX 1948—Grub.); ***Cent. Khalkha*** (Dzhargalante river basin, 47°N. lat., 104–105°E long., marshy meadows, 11 VIII and 15 IX 1925—Krasch. and Zam.), ***East. Mong., Depr. Lakes*** (Kharikhiry river stream 3–4 km south of Ulangom, marshy meadows, 28 VII 1945—Yun.).

IIA. Junggar: ***Zaisan*** (Belezek river lower course, tugai, 18 VI 1914—Schischk.).

General distribution: Arct., Europe, Asia Minor, Caucasus, West. Sib., East. Sib., Nor. Mong., Nor. Amer.

19. **J. articulatus** L. Sp. pl. (1753) 465; Fl. Kirgiz. 3 (1951) 21; Fl. Kazakhst. 2 (1958) 99; Fl. Tadzh. 2 (1963) 173; Ikonnikov, Opred. rast. Pamira (Key to Plants of Pamir) (1963) 84. —*J. lampocarpus* Ehrh. Calam. (about 1791) No. 126; Davies in Trans. Linn. Soc. London 10 (1811) 13; Franch. Pl. David. 1 (1884) 311; Buchenau in Pflanzenreich [IV, 36], 25 (1906) 217; Krylov, Fl. Zap. Sib. 3 (1929) 576; ?Pampanini, Fl. Carac. (1930) 86; V. Krecz. and Gontsch. in Fl. SSSR, 3 (1935) 538; Persson in Bot. Notis. (1938) 277; Bohlin in Rep. Sci. Exped. North-West. Prov. China S. Hedin, 11, 3 (1949) 28. —*J. acutiflorus* auct. non Ehrh.: Krylov, l.c. 574. **Ic.:** Fl. SSSR, 3, Plate XXX, figs. 1 and 1a.

Described from Europe. Type in London (Linn.) Plate VII, fig. 10.

In marshy and wet short-grass meadows along banks of brooks and rivers, near springs, along irrigation ditches, meadows in floodplains and marshes on sasa solonchak; from foothills to 2800 m above sea level.

IA. Mongolia: ***Mong. Alt.*** (Bulugun river floodplain at inflow of Ulyaste-Gol into it, cereal grass meadows, 20 VII 1947—Yun.); ***East. Mong.*** (right bank of Huang He river below Khěkou town, 7 VIII 1884—Pot.); ***Alash. Gobi*** (40 km nor. of Yunchan town, Nin-yanlu settlement, wet meadow in valley, 1 VII 1958—Petr.), ***Khesi*** (Tszyutsyuan', No. 0011, 2 VII 1956—Ching; "Edsen-Gol, Bayan-bogdo, 1933, D. Hummel"—Bohlin, l.c.).

IB. Kashgar: ***Nor., West.*** (Yangigisar region, 27 V 1909—Divn.; near Yangishar, along irrigation ditches, 25 VII 1929—Pop.; Gez-Darya river valley 100 km from Kashgar, hot spring, in water, 15 VI 1957—Yun.; Bulunkul'-Upal, 2800 m, No. 00356, 15 XI 1959—Lee et al.; "Kashghar, about 1330 m, 18 IV 1933" Persson, l.c.); ***East.*** (in Turfan near water, 100 m, No. 5512, 1 VI; Turfan, Yarkhu, near water, 100 m, No. 5565, 9 VI; nor.-east of Toksun, marsh, 300 m, No. 7317, 19 VI; Nyutsugun to Tsagan-tunge in Khomot, No. 7711, 15 VIII 1958—Lee and Chu; south. border of Khami oasis, Bugas, along springs, 480 m, 18 VIII 1895—Rob.).

IIA. Junggar: ***Altay Region*** (Fuyun'-Ukagou, No. 2267, 20 VIII; Altay, No. 2580, 28 VIII 1956—Ching); ***Jung. Alt.*** (nor.-west. border of Dzhair mountain range, 24 km nor.-west of Toli settlement, Modun-obo brooks, near water, 5 VIII 1957—Yun.); ***Tien Shan, Jung. Gobi*** (nor.: Barbagai-Burchum, No. 2937, 11 IX 1956—Ching; south.: Tsitai, Nan'khu, No. 0668, 26 VII 1956—Ching; Manas river basin, 2 km nor. of Kuitun settlement, No. 1131, 29 VI; in Shikhetsza region, No. 1171, 2 VII 1957—Kuan; 3–4 km nor. of St. Kuitun settlement, sedge-herb meadow, 6 VII; 3–4 km east of St. Kuitun settlement, Khonkado area, marshy meadow, 7 VIII 1957—Yun.); ***Balkh.-Alak.*** (around Dachen town, in marsh, No. 2846, 10 VIII 1957—Kuan).

IIIA. Qinghai: ***Nanshan*** (Lovachen town, along bank of Xining river, 26 VII 1908—Czet.).

General distribution: Aral-Casp., Balkh. Region, Jung.-Tarb., Nor. and Cent. Tien Shan, East. Pam.; Arct. (Europe.—outside USSR, Asiat.—Olenek basin and Kolyma estuary), Europe, Near East, Caucasus, Mid. Asia, West. Sib., China (Dunbei, South-west.), Himalayas (west., Kashmir), North Amer. (nor.-east.), Africa (nor.).

20. **J. atratus** Krock. Fl. Siles. 1 (1787) 562; Buchenau in Pflanzenreich [IV, 36], 25 (1906) 210; Krylov, Fl. Zap. Sib. 3 (1929) 573; V. Krecz. and Gontsch. in Fl. SSSR, 3 (1935) 542; Fl. Kazakhst. 2 (1958) 99. —**Ic.:** Fl. SSSR, 3, Plate XXX, fig. 5.

Described from Europe (Silesia). Type ?

IIA. Junggar: ***Jung. Gobi*** (nor.: south of Shara-Sume, Barbagai, No. 3878, 8 IX 1956—Ching).

General distribution: Aral-Casp., Balkh. Region, Jung.-Tarb.; Europe, Caucasus, West. Sib. (south.), East. Sib. (south.).

***J. leschenaultii** J. Gay in Laharpe, Monogr. Jancac. (1827) 137; V. Krecz. and Gontsch. in Fl. SSSR, 3 (1935) 540. —*J. prismatocarpus* var. *leschenaultii* (J. Gay) Buchenau in Pflanzenreich [IV, 36], 25 (1906) 180. —*J. prismatocarpus* auct. non R. Br.: Forbes and Hemsley, Index Fl. Sin. 3 (1905) 165; Chen and Chou, Rast. pokrov r. Sulekhe (Plant Cover of Sulekhe River) (1957) 91.

Described from southern India (Nilgiri hills). Type ?

IA. Mongolia: ***Khesi*** ("Sulekhe river"—Chen and Chou, l.c.).

General distribution: Far East, China (North, Cent., South-west., East., South., Hainan, Taiwan), Japan, Korea, Indo-Malay.

21. **J. turczaninowii** (Buchenau) V. Krecz. in Fl. SSSR, 3 (1935) 539; Kitag. Lin. Fl. Mansh. (1939) 129.

Described from East. Siberia. Type in Leningrad.

In marshy meadows, along edges of puddles.

IA. Mongolia. ***Cis-Hing., East. Mong.*** (right bank of Huang He river below Khekou town, in marshy meadows, 7 VIII 1884—Pot.); ***Ordos*** (Huang He river valley, marshy meadows, 9 VIII 1871—Przew.; 10 km south-west of Ushin town, on edge of puddle, 4 VIII; 50 km south of Dzhasak town, meadow with puddle among dune sands, 17 VIII 1957—Petr.).

General distribution: East. Sib. (south. Transbaikal—Nerchinsk), *North Mong.* (Hang., Mong.-Daur.), China (Dunbei).

Note. This species, very close to *J. alpinus* Vill., differs from it in very short anthers and slightly more elongated perianth segments, of which inner sometimes slightly longer than outer. The latter feature is particularly characteristic of plants in the collection of M.P. Petrov and Ordos.

Subgenus **Juncotypus** (Dum.) Nakai

22. **J. brachytepalus** V. Krecz. et Gontsch. in Fl. SSSR, 3 (1935) 630, 547; Fl. Kirgiz. 3 (1951) 13; Fl. Kazakhst. 2 (1958) 99; Fl. Tadzh. 2 (1963) 174.

—*J. glaucus* auct. non Ehrh.: Forbes and Hemsley, Index Fl. Sin. 3 (1905) 164; Buchenau in Pflanzenreich [IV, 36], 25 (1906) 133, p.p. —**Ic.:** Fl. Tadzh. 2, Plate XXXIX, figs. 4 and 5.

Described from Kazakhstan. Type in Leningrad. Plate VIII, fig. 11.

In grassy marshes, along banks of rivers; up to 1500 m above sea level.

IB. Kashgar: ***East.*** (nor.-east of Toksun, in marsh, 300 m, No. 7321, 19 VI 1958—Lee and Chu).

IIA. Junggar: ***Tien Shan*** (Piluchi, VI 1877; Nilki, 1500 m, VI 1879—A. Reg.; upper course of Kunges river, 1050 m, 29 VI 1877—Przew.); ***Dzhark.*** (Khoyur-Sumun south of Kul'dzha, 28 V 1879—A. Reg.); ***Balkh.-Alak.*** (around Dachen town [Chuguchak], in marsh, No. 2827, 10 VIII 1957—Kuan).

General distribution: Aral.-Casp., Balkh. Region, Jung.-Tarb., Nor. and Cent. Tien Shan, ?East. Pam.; Near East, Mid. Asia (Turkmenia, West Tien Shan, Pamiroalay), China (Nor.-west.).

Note. Very close to *J. inflexus* L., from which it differs mainly in very short bract leaf as well as very short perianth segments.

***J. decipiens** (Buchenau) Nakai, Rep. Veg. Kamikoti (1928) 35; V. Krecz. and Gontsch. in Fl. SSSR, 3 (1935) 551; Kitag. Lin. Fl. Mansh. (1939) 127. —*J. effusus* var. *decipiens* Buchenau in Pflanzenreich [IV, 36], 25 (1906) 136. —*J. effusus* auct. non L.: Forbes and Hemsley, Index Fl. Sin. 3 (1905) 163; Chen and Chou, Rast. pokrov r. Sulekhe (Plant Cover of Sulekhe River) (1957) 91 **Ic.:** Fl. SSSR, 3, Plate XXIX, fig. 9.

Described from Japan. Type in Tokyo (?).

IA. Mongolia: ***Khesi*** ("Sulekhe river"—Chen and Chou, l.c.).

General distribution: Far East (south.), China (Dunbei, South-west., South., Taiwan), Korea, Japan.

Note. It is possible that reports about the occurrence of *J. decipiens* (Buchenau) Nakai (= *J. effusus* auct.) within Central Asia are erroneous and refer to some other species.

23. **J. filiformis** L. Sp. pl. (1753) 326; Buchenau in Pflanzenrich, [IV, 36], 25 (1906) 127; Krylov, Fl. Zap. Sib. 3 (1929) 568; V. Krecz. and Gontsch. in Fl. SSSR, 3 (1935) 552; Fl. Kazakhst. 2 (1958) 100. —**Ic.:** Fl. SSSR, 3, Plate XXIX, figs. 1 and 12.

Described from Europe (Sweden). Type in London (Linn.).

IIA. Junggar: ***Altay Region*** (Qinhe—Kun'tai, 2650 m, No. 0988, 9 VIII; in Fuyun' [Koktogoi] region, 2500 m, No. 1907, 17 VIII 1956—Ching).

General distribution: Aral-Casp.; Arct., Europe, Caucasus, West. Sib., East. Sib., Far East, North Amer., ?South Amer. (Patagonia).

2. **Luzula** DC.

in Lam. and DC. Fl. Franc. 1 (1805) 158.

1. Inflorescence spicate, often drooping. Bract leaf shorter than inflorescence. Bracts elongated, almost as long as flowers, highly fimbriate. Leaves not thickened towards tip.......3. **L. spicata** (L.) DC.

\+ Inflorescence umbellate or capitate, not drooping. Bract leaf equal to inflorescence or longer. Bracts not fimbriate, shorter than flowers. Leaves thickened toward tip..........2.

2. Perianth segments largely equal, dark brown. Inflorescence umbellate or capitate ....2. **L. sibirica** V. Krecz.
+ Inner perianth segments shorter than outer; all of them light or pale rust coloured. Inflorescence umbellate.......................................
....1. **L. pallescens** (Wahl.) Bess.

1. **L. pallescens** (Wahl). Bess. Enum. Pl. Volhyn. (1822) 15; Krylov, Fl. Zap. Sib. 3 (1929) 554; V. Krecz. and Gontsch. in Fl. SSSR, 3 (1935) 576; Kitag. Fl. Mansh. (1939) 129; Fl. Kirgiz. 3 (1951) 21; Grubov, Consp. fl. MNR (1955) 89; Fl. Kazakhst. 2 (1958) 102 —*L. campestris* var. *pallescens* Wahl. Fl. Suec. 1 (1824) 218; Buchenau in Pflanzenreich [IV, 36], 25 (1906) 88. —*Juncus pallescens* Wahl. Fl. Lapp. (1812) 87. —**Ic.:** Fl. Kazakhst. 2, Plate VIII, fig. 7.

Described from Sweden. Type in Uppsala.

In meadows and glades, along banks of rivers; in forest and subalpine belts, from 900 to 3000 m above sea level.

IA. Mongolia: ***Cis-Hing., Mong. Alt., East. Mong.***

IIA. Junggar: ***Jung. Alt.*** (Syaeda—Ven'tsyuan], 2700–2900 m, Nos. 1400 and 1410, 13 VIII 1957—Kuan); ***Tien Shan*** (Khanakhai hill, 1500–2100 m, 16 VI 1878—A. Reg.; Kash river, 900–1200 m, 3 VII; Borgaty, 1500–1800 m, 4 VII; Mengute, 2700 m, 2 VIII 1879—A. Reg.; Tekes river upper course, 3000 m, on river bed, 25 VI 1893—Rob.; nor. foothill of Narat mountain range, subalpine meadow, 7 VIII 1958—Yun.).

General distribution: Jung.-Tarb., Nor. and Cent. Tien Shan; Arct. (Europ. and Asiat.—stray finds), Europe, Asia Minor, Caucasus, West. Sib., East. Sib., Far East, North Mong. (Fore Hubs., Hent., Hang., Mong.-Daur.), China (Dunbei).

2. **L. sibirica** V. Krecz. in Fl. Zabaik. 2 (1931) 144; idem, Fl. SSSR, 3 (1935) 631, 574; Fl. Kirgiz. 3 (1951) 22; Grubov, Consp. fl. MNR (1955) 90; Fl. Kazakhst. 2 (1958) 102. —*L. multiflora* ssp. *asiatica* Kryl. et Serg. in Krylov, Fl. Zap. Sib. 3 (1929) 556. —*L. multiflora* auct. non Lej.: Kitag. Lin. Fl. Mansh. (1939) 129; Grubov, Consp. fl. MNR (1955) 89. —*L. sudetica* auct. non DC.: Kitag. l.c. 130.

Described from East. Siberia. Type in Leningrad.

In alpine and subalpine meadows, less often in upper forest boundary.

IA. Mongolia: ***Khobd.*** (Tszusylan, in forest, 12 and 13 VII; Kharkhiry river mouth, 24 VII 1879—Pot.); ***Mong. Alt.*** (Aksu river, slopes of glacier and moraine, 23 VII 1909—Sap.; Kharagaitu-daba, alpine meadow, 23 VII; Kharagaitu pass in upper Indertiin-Gol, alpine meadow 24 VII 1947—Yun.).

IIA. Junggar: ***Altay Region*** (Qinhe, Chzhunkhaitsza, No. 1388, 6 VIII 1956—Ching; in Timulbakhana region, 2450 m, No. 10740, 19 VII 1956—Lee et al.); ***Tien Shan*** (west. tributary of Aryslyn river, 2700 m, 8 VII; Aryslyn, 16 VII 1879—A. Reg.; Boro-Khoro mountain range, 2 km nor.-east of Sairam pass in Sairam lake basin, subalpine belt, border of spruce grove, 19 VIII 1957—Yun.; Borgate-Yakou, 2900 m, No. 1671, 30 VIII 1957—Kuan).

General distribution: Jung.-Tarb., North and Centr. Tien Shan; Arct. (Asiat.), West. Sib. (Altay), East. Sib. (south. and north Yakutia), Far East (south.), North Mong. (Fore Hubs., Hent., Hang. Mong.-Daur.), China (Dunbei).

3. **L. spicata** (L.) DC. in Lam. and DC. Fl. France, 3 (1805) 161; Buchenau in Pflanzenreich [IV, 36], 25 (1906) 73; Krylov, Fl. Zap. Sib. 3 (1929) 553;

Pampanini, Fl. Carac. (1930) 86; V. Krecz. and Gontsch. in Fl. SSSR 3 (1935) 570; Fl. Kirgiz. 3 (1951) 22; Grubov, Consp. fl. MNR (1955) 90; Fl. Kazakhst. 2 (1958) 101; Fl. Tadzh. 2 (1963) 176; Ikonnikov, Opred. rast. Pamira (Key to Plants of Pamir) (1963) 85. —*Juncus spicatus* L. Sp. pl. (1753) 330. **Ic.:** Fl. Tadzh. 2, Plate XL, fig. 1.

Described from Sweden. Type in London (Linn.).

In wet meadows in alpine belt.

IA. Mongolia: ***Khobd.*** (Kharkhiry river mouth, 24 VII 1879—Pot.); ***Mong. Alt.*** (Kharagaitu-daba pass, Indertiin-Gol river upper course, alpine meadow, 24 VII 1947—Yun.); ***Gobi-Alt.*** (Baga-Bogdo hills, fringes of vegetation, 30 VII 1895—Klem.).

IIA. Junggar: ***Altay Region*** (Fuyun' [Koktogoi]—Ukagou, 900 m, No. 1699, 11 VIII 1956—Ching; in Timulbakhana region, high-altitude belt 10703, 19 VII 1959—Lee et al.); ***Jung. Alt.*** (Ven'tsyuan', on slope, 2640 m, No. 2030, 25 VIII 1957—Kuan), ***Tien Shan*** (Muzart, 2700 m, 19 VIII 1877—A. Reg.; Mengut, 2700 m, 9 VII; Aryslyn, 3000–3300 m, 13 VII 1879—A. Reg; Manas river basin, Danu-Gol river upper course, high-altitude belt, Kobresia meadow, 21 VII 1957—Yun.; 6–7 km south of Danu, No. 498, 22 VII; 11 km east of Aksu on way to Chzhaosu, along roadside, No. 1579, 15 VIII 1957—Kuan).

General distribution: Jung.-Tarb., North and Centr. Tien Shan, East. Pam.; Arct. (Europ.), Europe (hills of Cent. and South. Europe, Khibiny, Nor. Urals, Carpathians), Caucasus, Mid. Asia (West. Pamiroalay), West. Sib. (south.), East. Sib. (south.), North Mong. (Fore Hubs., Hang.), Himalayas (west.), North Amer.

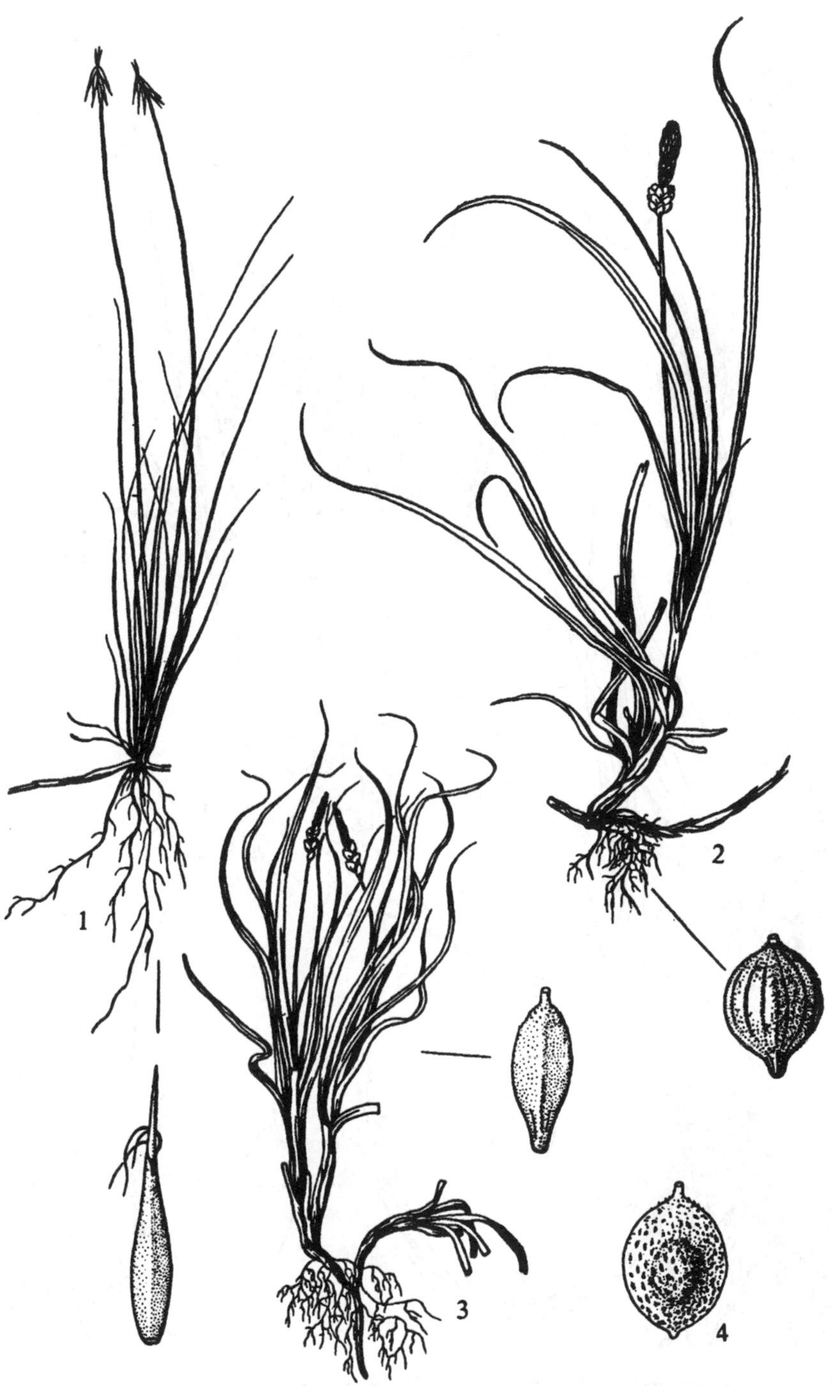

Plate I. *1—Carex microglochin* Wahl.; 2—C. *argunensis* Turcz. ex Trev.; 3—C. *rupestris* Bell. ex All.; 4—C. *kansuensis* Nelmes.

Plate II. *1—Carex melananthiformis* Litv.; 2—*C. melanantha* C. A. Mey.; 3—*C. moorcroftii* Falc. ex Boott; 4—*C. melanocephala* Turcz.

Plate III. 1—*Carex. relaxa* V. Krecz.; 2—*C. Ivanoviae* Egor.; 3—*C. turkestanica* Rgl.

Plate IV. *1—Carex lanceolata* var. *alaschanica* Egor.; *2—C. aneurocarpa* V. Krecz.; *3—C. duthiei* Clarke; *4—Kobresia tibetica* Maxim.

Plate V. *1—Carex alajica* Litv.; *2—C. przewalskii* Egor.; *3—C. lehmannii* Drej.

1
2
3
4

Plate VI. *1*—*Carex stenocarpa* Turcz. ex V. Krecz.; *2*—*C. macrogyna* Turcz. ex Steud.; *3*—*C. alexeenkoana* Litv.; *4*—*C. atrofusca* Schkuhr.

Plate VII. *1—Kobresia robusta* Maxim.; *2—K. filifolia* (Turcz.) Clarke; *3—K. humilis* (C.A. Mey.) Serg.; *4—K. stenocarpa* (Kar. et Kir.) Steud.; *5—K. simpliciuscula* (Wahl.) Mack.; *6—K. bellardii* (All.) Degl.; *7—K. capilliformis* Ivanova; *8—K. persica* Kük. et Bornm.; *9—K. smirnovii* Ivanova; *10—Juncus articulatus* L.

Plate VIII. *1—Eeleocharis argyrolepis* Kier. ex Bge.; *2—E. equisetiformis* (Meinsh.) B. Fedtsch.; *3—Juncus tibeticus* Egor.; *4—J. przewalskii* Buchenau; *5—J. leucomelas* Royle; *6—J. thomsonii* Buchenau; *7—J. triglumis* L.; *8—J. allioides* Franch.; *9—J. gerardii* Lois.; *10—J. salsuginosus* Turcz.; *11—J. brachytepalus* V. Krecz. et Gontsch.; *12—Blysmus rufus* (Huds.) Link; *13—B. compressus* (L.) Panz. ex Link; *14—B. sinocompressus* Tang et Wang; *15—Bolboschoenus planiculmis* (Fr. Schmidt) Egor.; *16—B. popovii* Egor.; *17—B. maritimus* (L.) Palla.

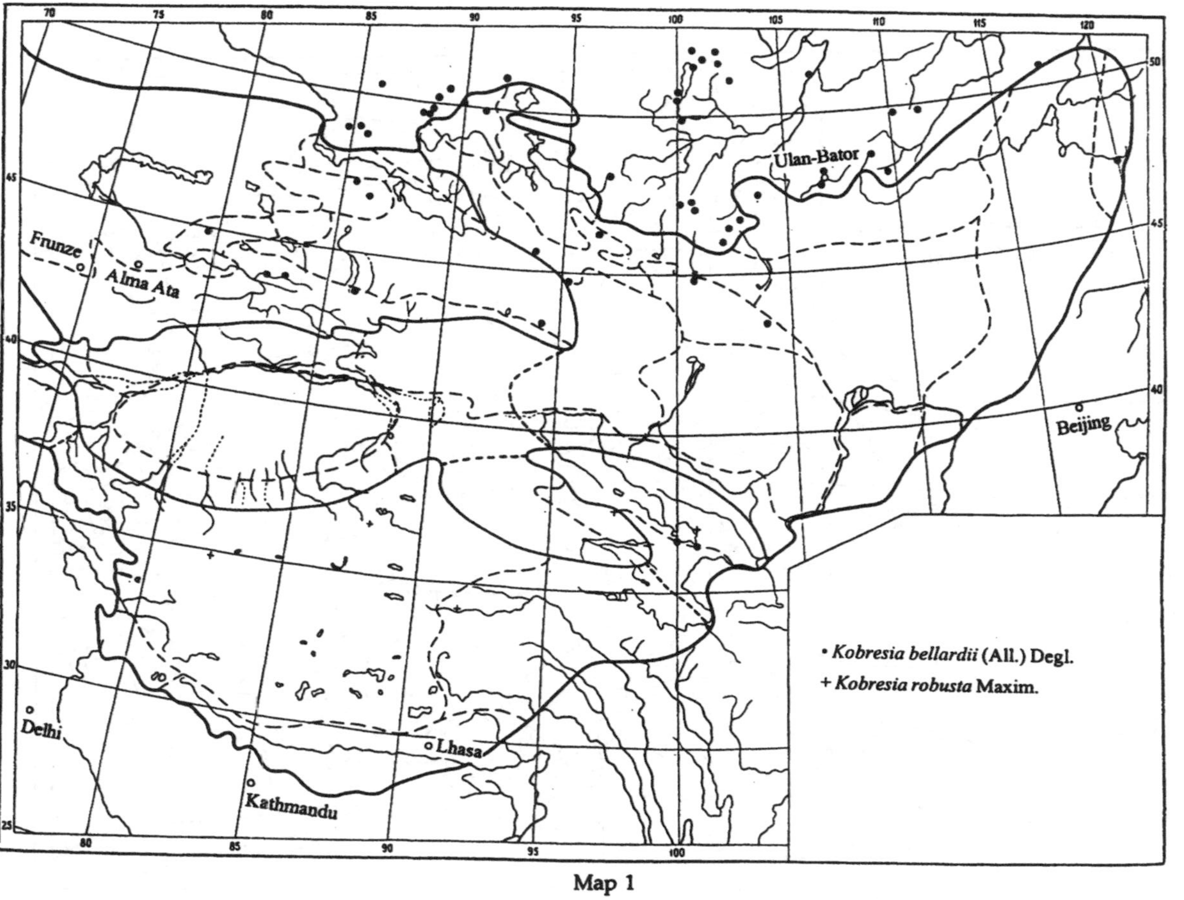
Frunze
Alma Ata
Ulan-Bator
Beijing
Delhi
Lhasa
Kathmandu
• *Kobresia bellardii* (All.) Degl.
+ *Kobresia robusta* Maxim.

Map 1

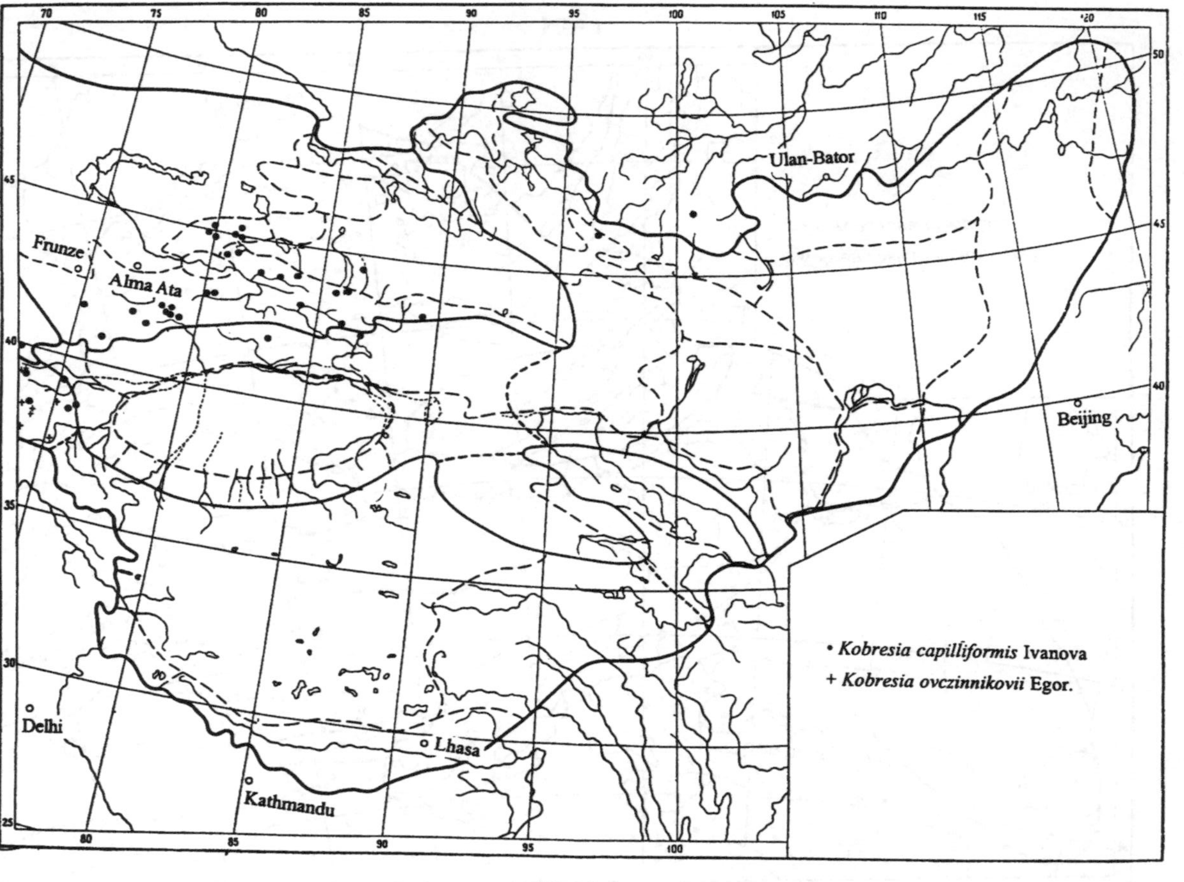

• Kobresia capilliformis Ivanova
+ Kobresia ovczinnikovii Egor.
Ulan-Bator
Beijing
Frunze
Alma Ata
Delhi
Lhasa
Kathmandu

Map 2

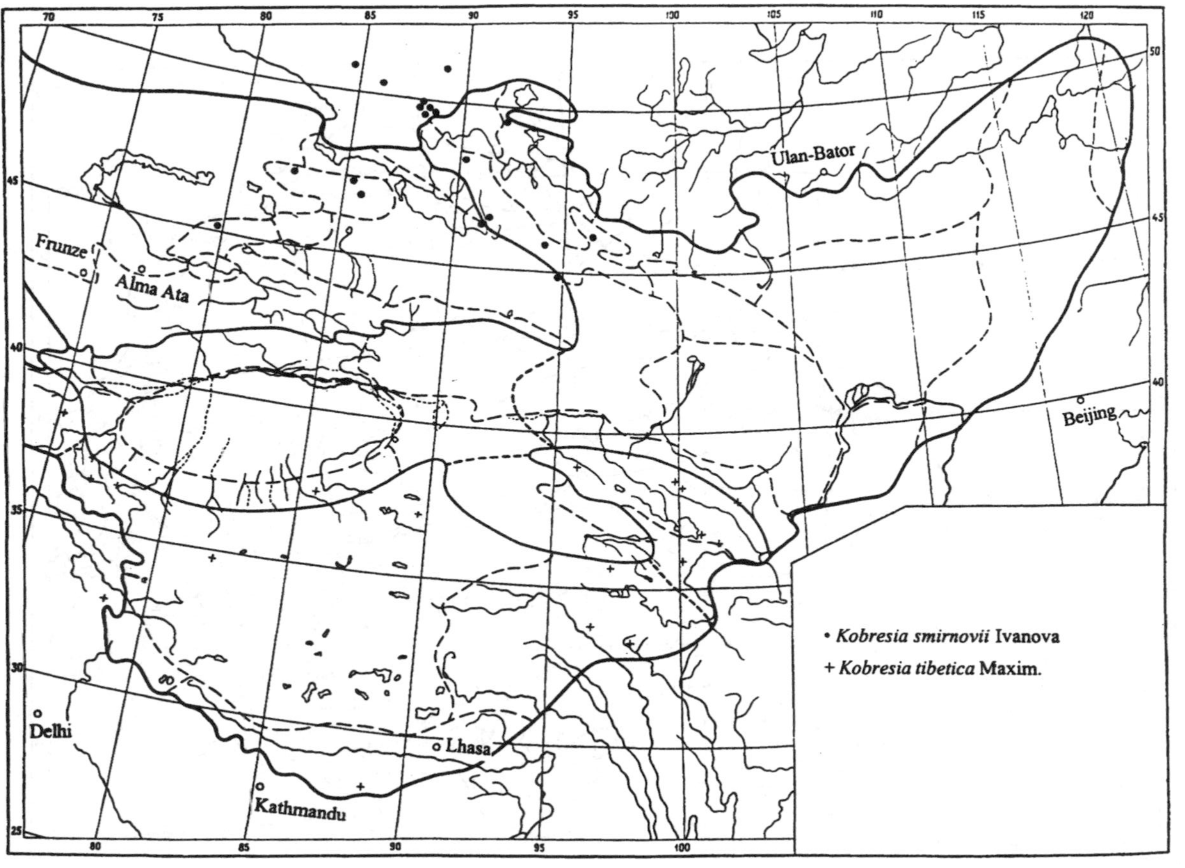
Ulan-Bator
Beijing
Frunze
Alma Ata
Delhi
Kathmandu
Lhasa
• *Kobresia smirnovii* Ivanova
+ *Kobresia tibetica* Maxim.

Map 3

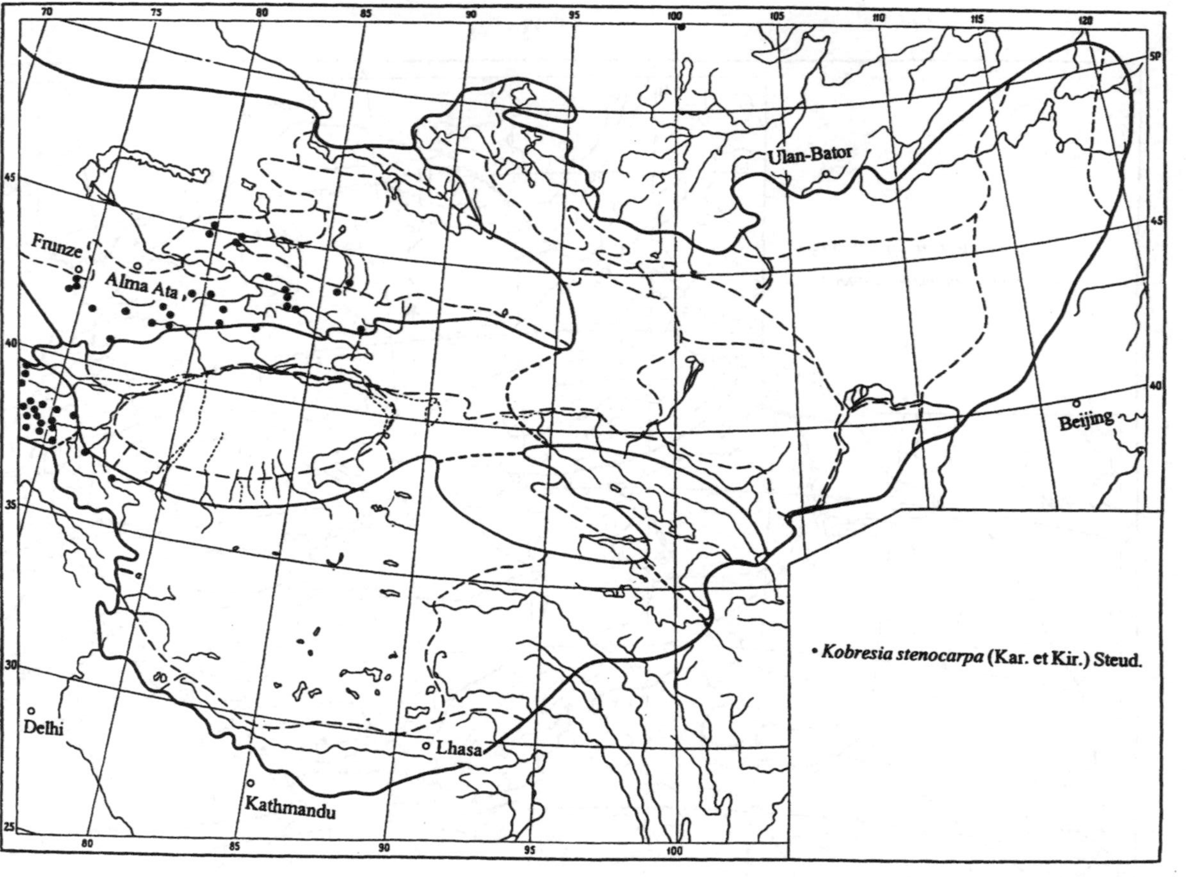
Ulan-Bator
Beijing
Frunze
Alma Ata
Delhi
Lhasa
Kathmandu
• *Kobresia stenocarpa* (Kar. et Kir.) Steud.

Map 4

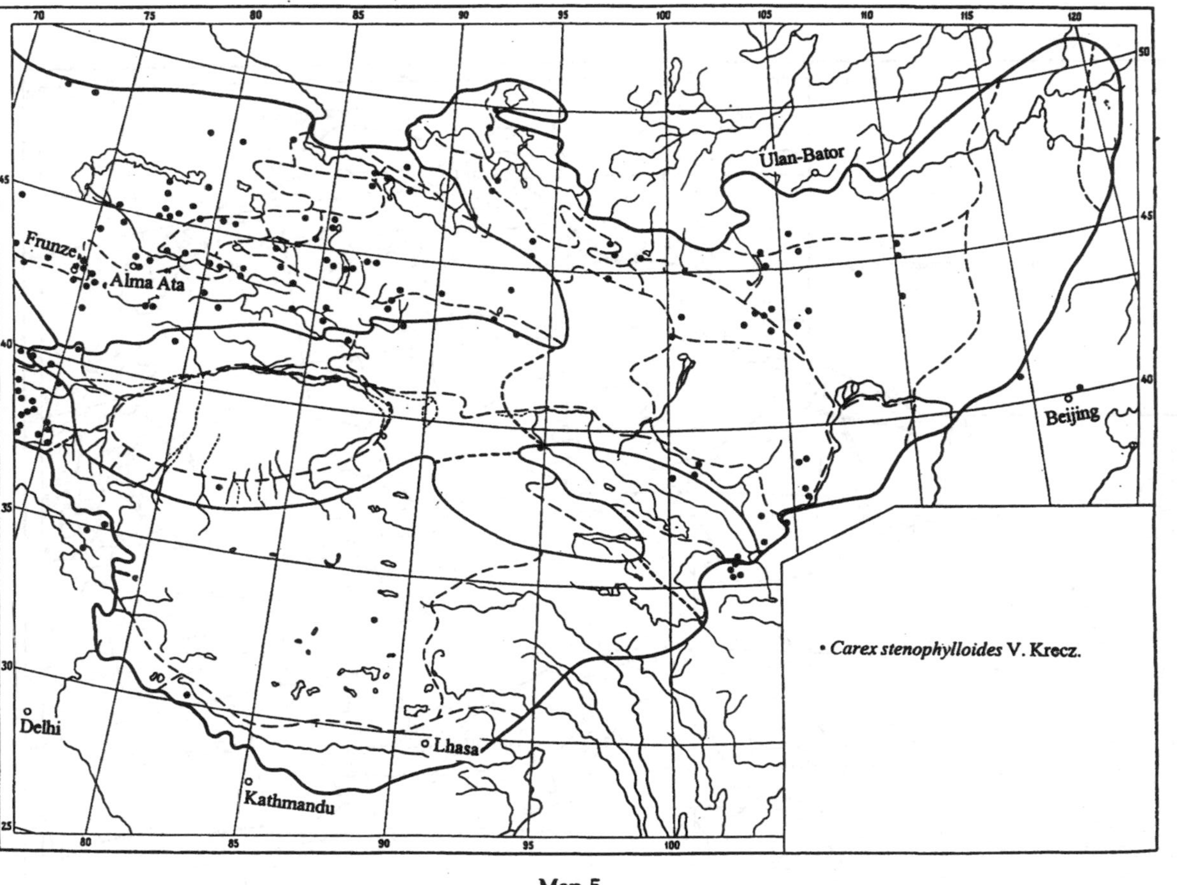

Ulan-Bator
Frunze
Alma Ata
Beijing
Delhi
Lhasa
Kathmandu
• Carex stenophylloides V. Krecz.

Map 5

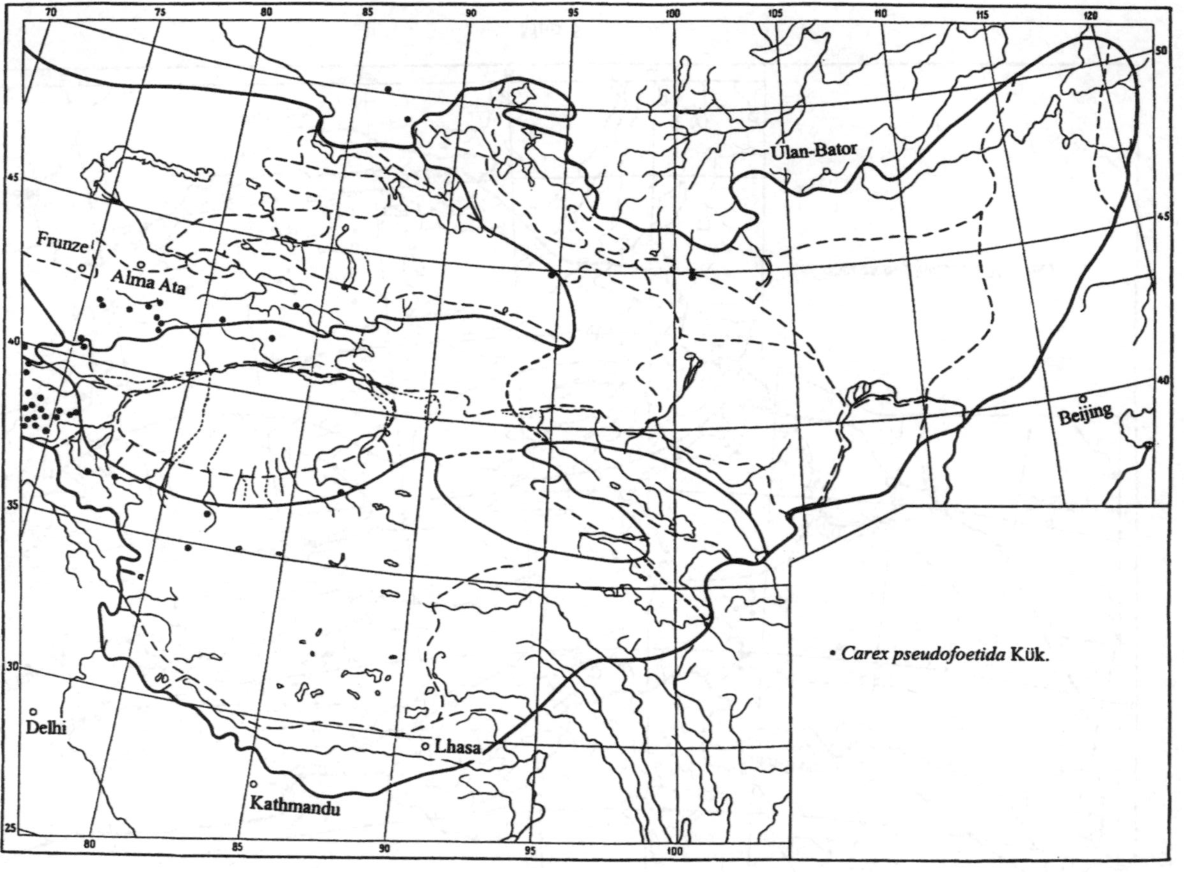
Ulan-Bator
Beijing
Frunze
Alma Ata
Delhi
Lhasa
Kathmandu
• *Carex pseudofoetida* Kük.

Map 6

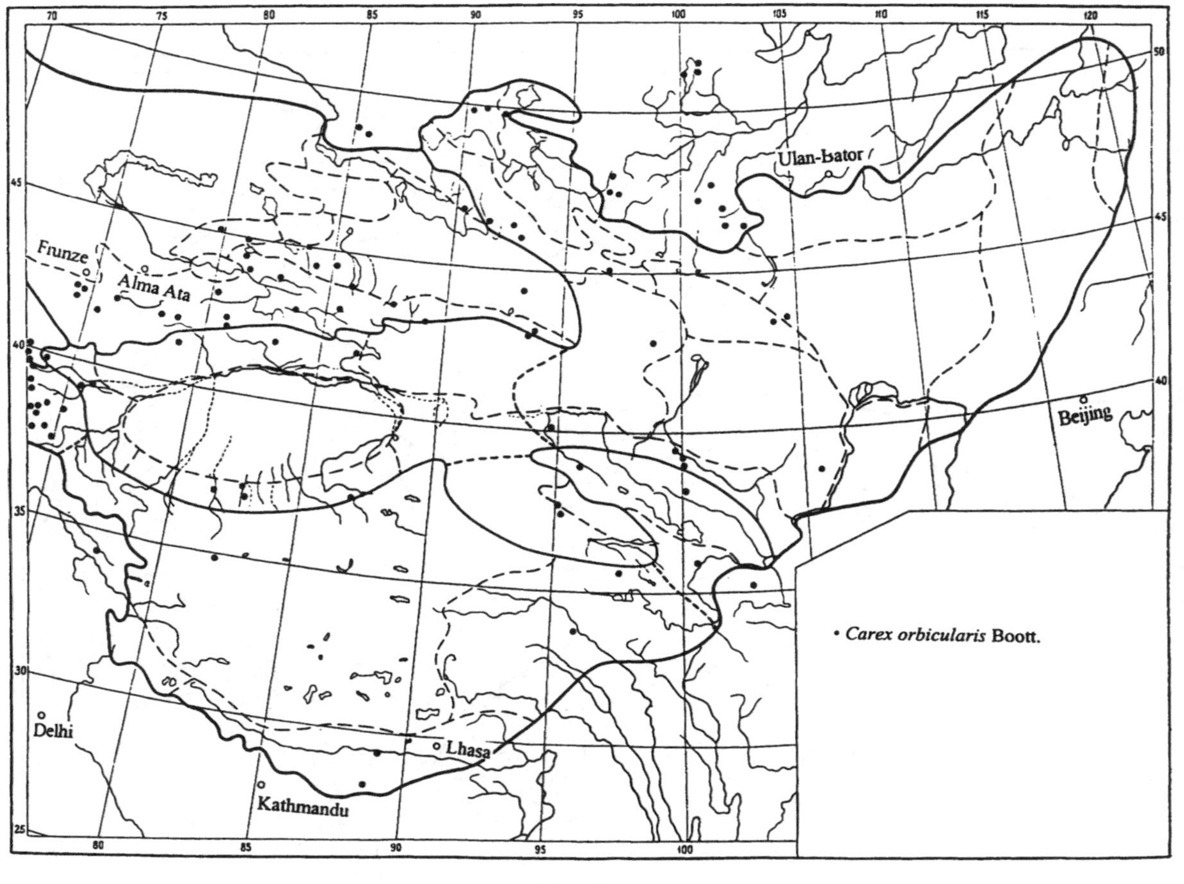
Ulan-Bator
Beijing
Frunze
Alma Ata
Lhasa
Delhi
Kathmandu
• *Carex orbicularis* Boott.

Map 7

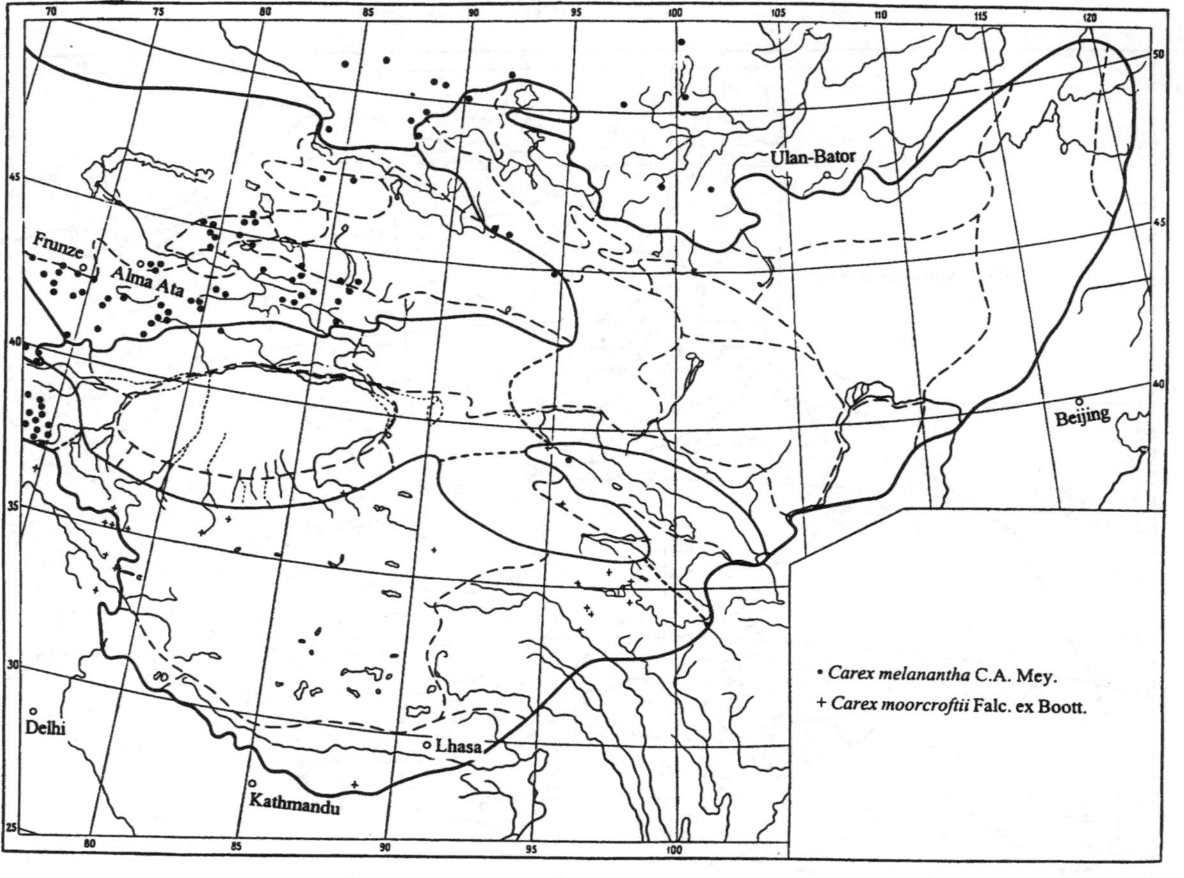

Ulan-Bator
Frunze
Alma Ata
Beijing
Delhi
Lhasa
Kathmandu
• Carex melanantha C.A. Mey.
+ Carex moorcroftii Falc. ex Boott.

Map 8

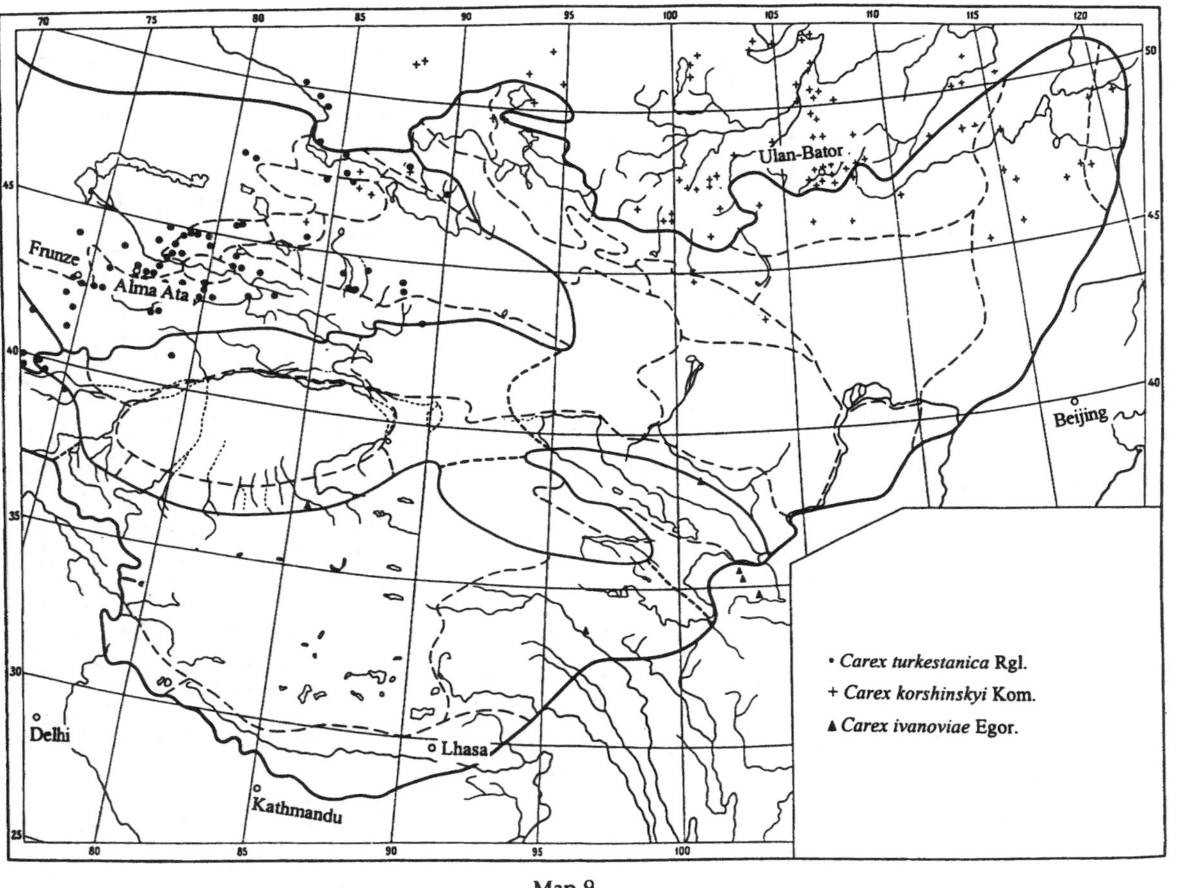

• Carex turkestanica Rgl.
+ Carex korshinskyi Kom.
▲ Carex ivanoviae Egor.
Ulan-Bator
Beijing
Frunze
Alma Ata
Delhi
Kathmandu
Lhasa

Map 9

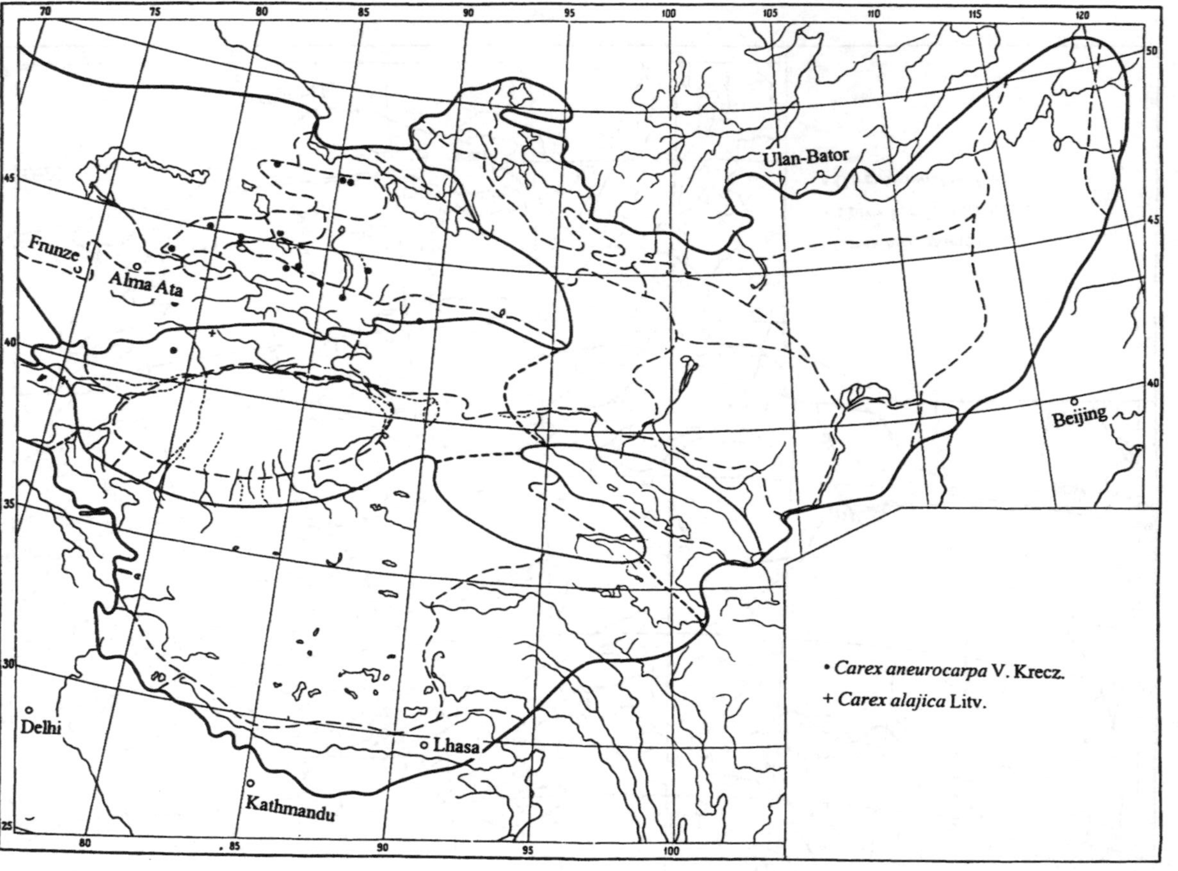
Ulan-Bator
Frunze
Alma Ata
Beijing
Delhi
Lhasa
Kathmandu
• Carex aneurocarpa V. Krecz.
+ Carex alajica Litv.

Map 10

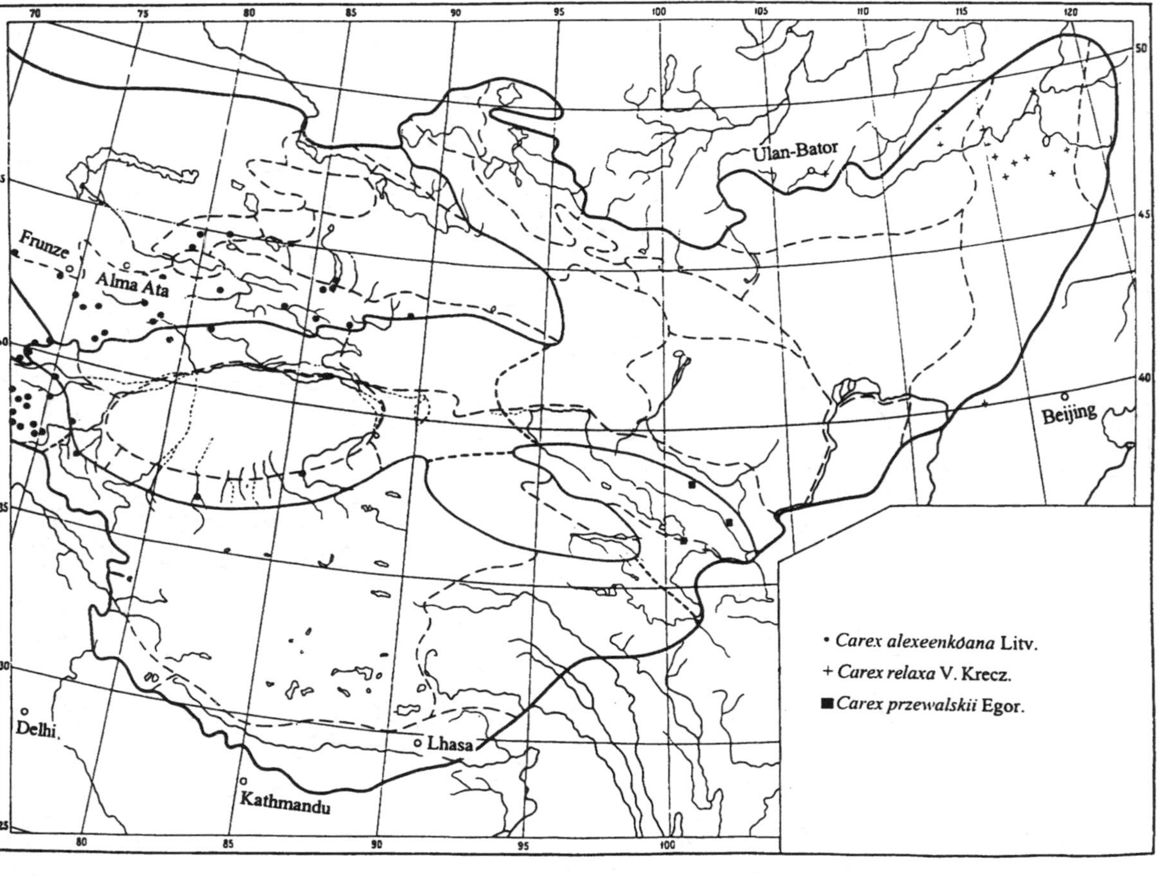

• Carex alexeenkoana Litv.
+ Carex relaxa V. Krecz.
■ Carex przewalskii Egor.
Ulan-Bator
Beijing
Frunze
Alma Ata
Lhasa
Kathmandu
Delhi

Map 11

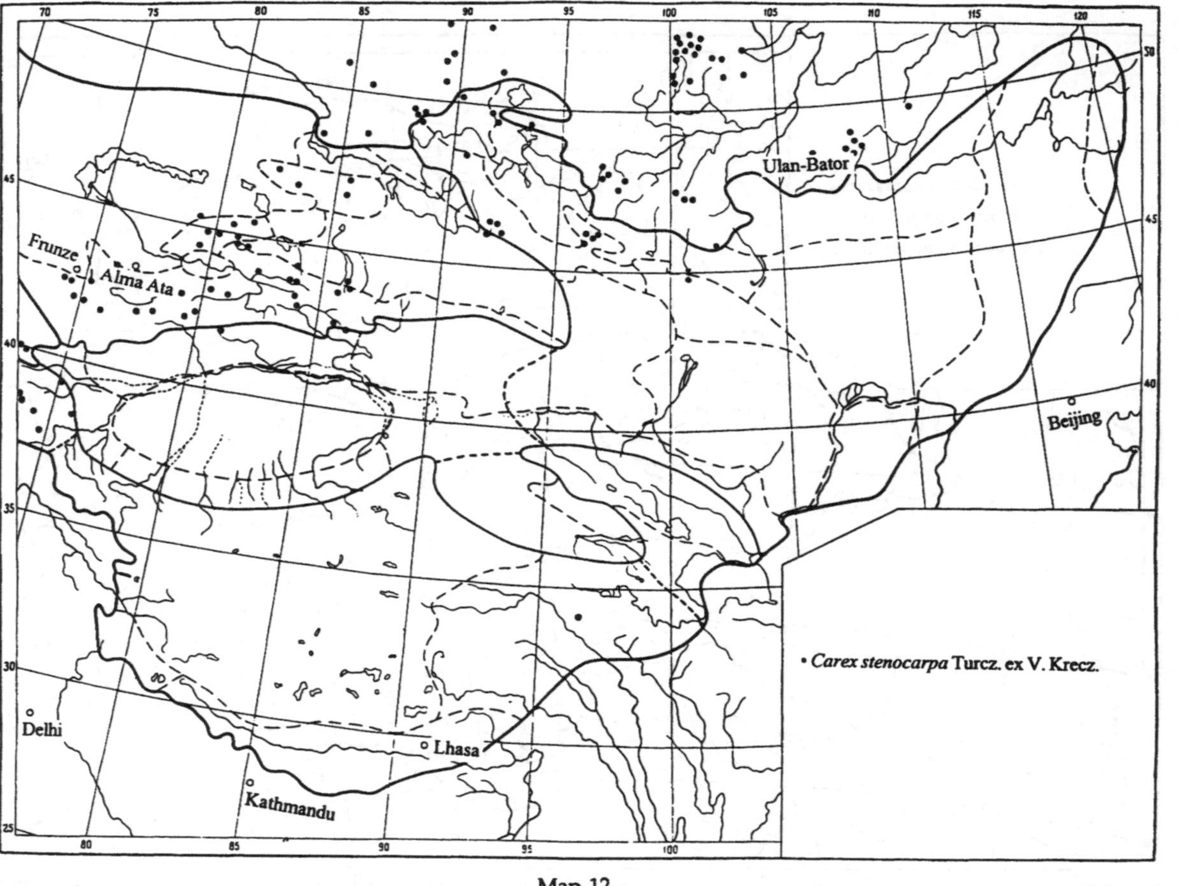
Ulan-Bator
Beijing
Frunze
Alma Ata
Delhi
Lhasa
Kathmandu
• *Carex stenocarpa* Turcz. ex V. Krecz.

Map 12

INDEX OF LATIN NAMES OF PLANTS

INDEX OF PLANT ILLUSTRATIONS

INDEX OF PLANT DISTRIBUTION RANGES

ADDENDA

Below is a listing of taxa which were found in the Mongolian Province and the Junggar-Turanian Province of the Central Asia after the publication of the Russian edition in 1967. The information on locating the taxa is contained in papers by many investigators (R. Kamelin, I. Gubanov et al., V. Novikov, V. Novikov and I. Gubanov, Z. Karamysheva and D. Banzragch, W. Hilbig and Z. Schamsran, E. Rachkovskaja and Ch. Sanchir, F. Jaeger et al., etc.). Recently these data were summarized by I. Gubanov in his book "Conspect flory Vneshney Mongolii" (1966), quoted below under each taxon. Central Asian regions, which add the geographic distribution of the taxa given in the Russian edition, are incorporated into the main English text.

It should be noted that some taxa from the cited below (Mariscus hamulosus, Eleocharis mitracarpa, Carex curaica ssp. pycnostachya and C. lachenalii) were already pointed out by me for Central Asia, but under different names.

Many herbarium specimens, on which the corresponding literature reports are based, were examined by author (Egorova).

Family **CYPERACEAE** Juss.

Eriophorum L.

E. mandshuricum Meinsh. Acta Horti Petrop. 18, 3 (1901) 268; Novoselova in Bot. zh. 78, 8 (1993) 87; Gubanov, Consp. fl. Vneshn. Mong. (1996) 28.

IA. Mongolia: *Mong. Alt., Cis-Hing.*

E. polystachion L. ssp. komarovii (V. Vassil.) Novosselova in Bot. zh. 79, 11 (1994) 87; Gubanov, Consp. fl. Vneshn. Mong. (1996) 28.—E. komarovii V. Vassil. in Bot. mater. Gerb. Bot. inst. AN SSSR. 8, 7 (1940) 102.—E. latifolium auct. non L.: Egorova in Rast. Centr. Asii, 3 (1967) 13.

IA. Mongolia: *Khobd., Mong. Alt., Cis-Hing., Cent. Khalkha, East. Mong.*

Note. The data on the occurrence E. latifolium L. within Central Asia given by me earlier (1967) are erroneous and refer to the above taxon.

E. scheuchzeri Hoppe ssp. altaicum (Meinsh.) Bondareva in Fl. Sib. 3 (1990) 15; Gubanov, Consp. fl. Vneshn. Mong. (1996) 28. —E. altaicum Meinsh. in Acta Horti Petrop. 18, 3 (1901) 267.

IA. Mongolia: *Khobd., Mong. Alt.*

Scirpus L.

S. radicans Schkuhr in Ann. Bot. Usteri, 2, 4 (1793) 49; Gubanov, Consp. fl. Vneshn. Mong. (1996) 28.

IA. Mongolia: ***East. Mong.***

Mariscus Vahl

M. hamulosus (M.B.) Hooper in Kew Bull. 26, 3 (1972) 578; Gubanov, Consp. fl. Vneshn. Mong. (1996) 28. —Cyperus hamulosus M.B. Fl. taur.-cauc. 1 (1808) 35. —Dichostylis hamulosa (M.B.) Nees in Linnaea, 9 (1834) 289; Egorova in Rast. Centr. Asii, 3 (1967) 24.

IA. Mongolia: ***Depr. Lakes.***
IIA. Junggar : ***Junggar Gobi.***

Eleocharis R. Br.

E. mamillata Lindb.f. in Doefler, Herb. Norm. 44 (1902) 108; Gubanov, Consp. fl. Vneshn. Mong. (1996) 28.

IA. Mongolia: ***Depr. Lakes.***

E. mitracarpa Steud. Syn Pl. Glum. 2 (1855) 77; Gubanov, Consp. fl. Vneshn. Mong. (1996) 28. —E. equisetiformis (Meinsh.) B. Fedtsch. Rast. Turkest. (1915) 165; Egorova in Rast. Centr. Asii, 3 (1967) 26.

IA. Mongolia: ***Mong. Alt.***

Kobresia Willd.

K. simpliciuscula (Wahl.) Mackenz. ssp. subholarctica Egor. in Nov. sist. vyssh. rast. 20 (1983), 83, ("subgolarctica"); Gubanov, Consp. fl. Vneshn. Mong. (1996) 28. —K. subholarctica (Egor.) Egor. in Bot. zh. 76, 12 (1991) 1736.

IA. Mongolia: ***Mong. Alt.*** (Munkh-Khairkhan mountain range).

Carex L.

C. arnelllii Christ in Scheutz, Kungl. Sv. Vet.-Akad. Handl. 22, 10 (1888) 177; Gubanov, Consp. fl. Vneshn. Mong. (1996) 25.

IA. Mongolia: ***Khobd., East. Mong.***

C. curaica Kunth ssp. pycnostachya (Kar. et Kir.) Egor. in Nov. sist. vyssh. rast. 10 (1973) 104; Gubanov, Consp. fl. Vneshn. Mong. (1996) 25. —C. pycnostachya Kar. et Kir. in Bull. Soc. natur. Moscou, 15, 3 (1842) 522.

IA. Mongolia: ***Depr. Lakes.***
IIA. Junggar: ***Junggar Gobi.***

C. diandra Schrank, Cent. Bot. Anmerk. (1781) 57; Gubanov, Consp. fl. Vneshn. Mong. (1996) 26.

IA. Mongolia: ***Cis-Hing., East. Mong., Depr. Lakes.***

C. **globularis** L. Sp. pl. (1753) 976; Gubanov, Consp. fl. Vneshn. Mong. (1996) 26.

IA. Mongolia: ***Mong. Alt.***

C. **heterolepis** Bge. Enum. pl. Chin. bor. (1832) 69; Gubanov, Consp. fl. Vneshn. Mong. (1996) 26.

IA. Mongolia: ***Cis-Hing.***

C. **lachenalii** Schkuhr, Beschr. Abbild. Riedgr. (1801) 51, tab. Y, fig. 79, nom. cons. —C. bipartita auct. non All.: Gubanov, Consp. fl. Vneshn. Mong. (1996) 25.

IA. Mongolia: ***Mong. Alt.***
IIA. Junggar: ***Junggar Gobi.***

C. **media** R. Br. in Richards. App. VII Bot., in Franklin, Narr. Journ. Polar Sea (1823) 750; Gubanov, Consp. fl. Vneshn. Mong. (1996) 26. —C. angarae Steud. Syn Pl. Glum. 2 (1855) 190.

IA. Mongolia: ***Khobd., Mong. Alt., Depr. Lakes.***

C. **leporina** L. Sp. pl. (1753) 973. —C. ovalis Good. in Trans. Linn. Soc. London, 2 (1794) 148; Gubanov, Consp. fl. Vneshn. Mong. (1996) 26.

IA. Mongolia: ***Mong. Alt.***

C. **parallela** (Laest.) Sommerf. ssp. redowskiana (C.A. Mey.) Egor. in Nov. sist. vyssh. rast. 10 (1973) 104. —C. redowskiana C.A. Mey. in Mém. Sav. Étr. Pétersb. 1 (1831) 207; Gubanov, Consp. fl. Vneshn. Mong. (1996) 27.

IA. Mongolia: ***Mong. Alt.***

C. **sordida** Van Heurck et Muell. Arg. in Van Heurck, Observ. Bot. 1 (1870) 33; Gubanov, Consp. fl. Vneshn. Mong. (1996) 27.

IA. Mongolia: ***East. Mong., Cis-Hing.***

C. **tomentosa** L. Mantissa (1767) 123; Gubanov, Consp. fl. Vneshn. Mong. (1996) 27.

IA. Mongolia: ***Cent. Khalkha.***

Family **LEMNACEAE** Gray

Lemna L.

L. turionifera Landolt in Aquatic Bot. 1 (1975) 355; Gubanov, Consp. fl. Vneshn. Mong. (1996) 29.

IA. Mongolia: ***East. Mong., Val. Lakes.***

Spirodela Schleid.

S. polyrhiza (L.) Schleid. in Linnaea, 13 (1839) 329; Gubanov, Consp. fl. Vneshn. Mong. (1996) 29. —Lemna polyrhiza L. Sp. pl. (1753) 970.

IA. Mongolia: ***Cis-Hing.***

Family **JUNCACEAE** Juss.

Juncus L.

J. alpinoarticulatus Chaix ssp. fischerianus (Turcz. ex V. Krecz.) Hämet-Ahti in Mem. Soc. Fauna Fl. Fenn. 56 (1980) 97; Gubanov, Consp. fl. Vneshn. Mong. (1996) 29. —J. fishcherianus Turcz. ex V. Krecz. Fl. Zabaik. 2 (1931) 142.

IA. Mongolia: ***Khobd., Mong. Alt., Cis-Hing., Cent. Khalkha, East. Mong., Depr. Lakes.***
IIA. Junggar: ***Junggar Gobi.***

J. ambiguus Guss. Fl. Sicul. Prodr. 1 (1827) 425. —J. bufonius L. ssp. ambiguus (Guss.) Schinz. et Thell. Fl. Schweiz., ed. 4 (1923) 126; Gubanov, Consp. fl. Vneshn. Mong. (1996) 29.

IA. Mongolia: ***Khobd., Mong. Alt., East. Mong., Depr. Lakes, Val. Lakes, Gobi Alt.***
IIA. Junggar: ***Junggar Gobi.***

J. biglumis L. Sp. pl. (1753) 328; Gubanov, Consp. fl. Vneshn. Mong. (1996) 29.

IA. Mongolia: ***Khobd., Mong. Alt.***

J. bufonius L. ssp. turkestanicus (V. Krecz. et Gontsch.) Novikov in Nov. sist. vyssh. rast. 24 (1987) 60; Gubanov, Consp. fl. Vneshn. Mong. (1996) 29.—J. turkestanicus V. Krecz. et Gontsch. in F1. SSSR, 3 (1935) 625.

IA. Mongolia: ***Mong. Alt., Cis-Hing., East. Mong., Depr. Lakes.***
IIA. Junggar: ***Junggar Gobi.***

J. gracillimus (Buchenau) V. Krecz. et Gontsch. in F1. SSSR, 3 (1935) 528; Gubanov, Consp. fl. Vneshn. Mong. (1996) 29.—J. compressus Jacq. var. gracillimus Buchenau in Pflanzenreich, [IV, 36] 25 (1906) 112.

IA. Mongolia: ***Mong. Alt., Cis-Hing.***

J. leucochlamys Zing. et V. Krecz. in F1. Zabaik. 2 (1931) 141; Gubanov, Consp. fl. Vneshn. Mong. (1996) 29.

IA. Mongolia: ***Khobd., Mong. Alt., East. Mong.***

J. orchonicus Novikov in Byull. Mosk. obshch. ispyt. prir. otdel. biol. 90, 5 (1981) 110; Gubanov, Consp. fl. Vneshn. Mong. (1996) 29.

IA. Mongolia: ***Cis-Hing., Cen. Khalkha, East. Mong., Depr. Lakes.***

J. triceps Rostk. Monogr. Junci (1801) 48; Gubanov, Consp. fl. Vneshn. Mong. (1996) 29.—J. castaneus Smith ssp. Triceps (Rostk.) Noikov in Nov. sist. vyssh. rast. 15 (1979) 92.

IA. Mongolia: ***Khobd., Mong. Alt.***

J. vvedenskyi V. Krecz. in Byull. Sredneaz. gos. univ. 21 (1935) 176 Gubanov, Consp. fl. Vneshn. Mong. (1996) 30.

IA. Mongolia: ***Khobd., Mong. Alt., Depr. Lakes, Val. Lakes, Gobi Alt.***
IIA. Junggar: ***Junggar Gobi.***

Luzula DC.

L. confusa Lindeb. in Bot. Not. (Lund) (1855) 9; Gubanov, Consp. fl. Vneshn. Mong. (1996) 30.

IA. Mongolia: ***Khobd., Mong. Alt.***

L. multiflora (Ehrh.) Lej. ssp. frigida (Buchenau) V. Krecz. in Zh. Russk. Bot. obshch. 12, 4 (1928) 490; Gubanov, Consp. fl. Vneshn. Mong. (1996) 30.—L. campestris (L.) DC. var. in Österr. Bot. Zeitschr. (1898) 284.—L. frigida (Buchenau) Sam. ex Lindm. Sv. Fanerogamfl. (1918) 161.

IA. Mongolia: ***Mong. Alt.***

L. pallidula Kirschner in Taxon, 39, 1 (1990) 110;
Gubanov, Consp. fl. Vneshn. Mong. (1996) 30. —L. pallescens auct. non (Wahl.) Bess.: Egorova in Rast. Centr. Asii, 3 (1967) 103 et auct. mult.

IA. Mongolia: ***Mong. Alt., Cis-Hing., East. Mong.***

L. parviflora (Ehrh.) Desv. in Journ. Bot. Appl. (Paris) 1 (1808) 144; Gubanov, Consp. fl. Vneshn. Mong. (1996) 30. —Juncus parviflorus Ehrh. Beitr. Naturk. 6 (1791) 139.

IA. Mongolia: ***Khobd., Mong. Alt.***

L. spicata (L.) DC. ssp. mongolica Novikov in Nov. sist. vyssh. rast. 26 (1989) 34; Gubanov, Consp. fl. Vneshn. Mong. (1996) 30.

IA. Mongolia: ***Khobd., Mong. Alt., Gobi Alt.***

T.V. Egorova
15 May 2000